AF358684

Spectroscopy-Based Biosensors

Spectroscopy-Based Biosensors

Editors

Annalisa De Girolamo
Vincenzo Lippolis
Chris Maragos

MDPI • Basel • Beijing • Wuhan • Barcelona • Belgrade • Manchester • Tokyo • Cluj • Tianjin

Editors
Annalisa De Girolamo
National Research Council of
Italy (CNR)
Italy

Vincenzo Lippolis
National Research Council of
Italy (CNR)
Italy

Chris Maragos
United States Department
of Agriculture
USA

Editorial Office
MDPI
St. Alban-Anlage 66
4052 Basel, Switzerland

This is a reprint of articles from the Special Issue published online in the open access journal *Biosensors* (ISSN 2079-6374) (available at: https://www.mdpi.com/journal/biosensors/special_issues/spectroscopy_biosensors).

For citation purposes, cite each article independently as indicated on the article page online and as indicated below:

LastName, A.A.; LastName, B.B.; LastName, C.C. Article Title. *Journal Name* **Year**, *Volume Number*, Page Range.

ISBN 978-3-0365-1002-6 (Hbk)
ISBN 978-3-0365-1003-3 (PDF)

Contents

About the Editors

Annalisa De Girolamo is a Senior Researcher at the Institute of Sciences of Food Production, National Research Council of Italy (CNR) (Bari, Italy). She has more than twenty-years of experience in the field of food safety and food authenticity with special interests in development and validation of conventional methods as well as rapid methods based on infrared spectroscopy combined with chemometrics. She has organized several collaborative studies for the validation of analytical methods, proficiency testing for the quality control of the laboratories, and related statistical data processing according to the AOAC/IUPAC guidelines and the ISO/IUPAC standards. She has also specific competence in selection of synthetic mediators (aptamers, molecular imprinted polymers) for several target molecules. Annalisa De Girolamo is involved in several national and EU-funded projects. She is author of 57 peer reviewed publications (h-index of 25, Scopus), 2 book chapters, 2 European reports. Tutor of several PhD and degree thesis on Food Safety. Guest editor of several ISI journals. Teacher at post-graduate masters. Referee for several International Scientific Journals. Speaker at several national and international Conferences/Symposiums.

Vincenzo Lippolis is a Senior Researcher at the Institute of Sciences of Food Production, National Research Council of Italy (Bari, Italy). He has more than fifteen-years of experience in the field of food safety and food authenticity with specific focus on the development and validation of targeted and non-targeted analytical methods based on fluorescence polarization immunoassays, lateral flow immunoassays, surface plasmon resonance, electronic noses and mass spectrometry. Vincenzo Lippolis is involved in several national and EU-funded projects. Vincenzo Lippolis is author of more than 40 peer reviewed publications (h-index of 14, Scopus), 1 patent and 2 book chapters. Supervisor of PhD thesis on Food Safety. Tutor of degree thesis in Analytical Chemistry. Teacher at post-graduate masters. Guest editor of several ISI journals. Referee for several International Scientific Journals. Speaker at several national and international Conferences/Symposiums.

Chris Maragos is a Lead Scientist in the Mycotoxin Prevention and Applied Microbiology Research Unit at the National Center for Agricultural Utilization Research (Peoria, Illinois, USA), a part of the U.S. Department of Agriculture - Agricultural Research Service. He has more than thirty years of experience in the field of food safety, with particular emphasis on natural toxins. His research interests involve the development of new technologies and materials for the detection of toxic secondary metabolites of fungi (mycotoxins) in commodities and foods. He is the author of 128 peer reviewed publications and book chapters (h-index of 38, Scopus). He is a member of the Editorial Boards of the journal Food Additives and Contaminants, Mycotoxin Research, and the World Mycotoxin Journal. He has served in various capacities within the AOAC International and is a Fellow of the AOAC International.

Preface to "Spectroscopy-Based Biosensors"

Biosensors are analytical devices capable of providing quantitative or semi-quantitative information by using a biological recognition element and a transducer. Depending upon the nature of the recognition element, different surface sensitive techniques can be applied to monitor these molecular interactions. In order to increase sensitivities and to lower detection limits down to even individual molecules, nanomaterials are promising candidates. This is possible due to the potential to immobilize more bioreceptor units at reduced volumes and their ability to act as transduction elements by themselves. Among such nanomaterials, gold nanoparticles, quantum dots, polymer nanoparticles, carbon nanotubes, nanodiamonds, and graphene are intensively studied. Biosensors provide rapid, real-time, accurate, and reliable information about the analyte under investigation and have been envisioned in a wide range of analytical applications, including medicine, food safety, bioprocessing, environmental/industrial monitoring, and electronics. A variety of biosensors, such as optical, spectroscopic, molecular, thermal, and piezoelectric, have been studied and applied in countless fields.

In this book, examples of spectroscopic and optical biosensors and immunoassays are presented. Furthermore, two comprehensive reviews on optical biosensors are included.

Annalisa De Girolamo, Vincenzo Lippolis, Chris Maragos
Editors

Review

Silver-Based Plasmonic Nanoparticles for and Their Use in Biosensing

Alexis Loiseau [1], Victoire Asila [2,†], Gabriel Boitel-Aullen [2,†], Mylan Lam [2,†], Michèle Salmain [3] and Souhir Boujday [1,*]

[1] Laboratoire de Réactivité de Surface (LRS), Sorbonne Université, CNRS, UMR 7197, 4 Place Jussieu, F-75005 Paris, France; alexis.loiseau@sorbonne-universite.fr

[2] Sorbonne Université, Faculté des Sciences et Ingénierie, Master de Chimie, Profil MatNanoBio, 4 Place Jussieu, F-75005 Paris, France; victoire.asila@upmc.fr (V.A.); g.boitelaullen@gmail.com (G.B.-A.); mylan951@hotmail.fr (M.L.)

[3] Institut Parisien de Chimie Moléculaire (IPCM), Sorbonne Université, CNRS, 4 Place Jussieu, F-75005 Paris, France; michele.salmain@sorbonne-universite.fr

* Correspondence: souhir.boujday@sorbonne-universite.fr; Tel.: +33-1-44-27-60-01

† Equally contributed to the manuscript.

Received: 3 May 2019; Accepted: 31 May 2019; Published: 10 June 2019

Abstract: The localized surface plasmon resonance (LSPR) property of metallic nanoparticles is widely exploited for chemical and biological sensing. Selective biosensing of molecules using functionalized nanoparticles has become a major research interdisciplinary area between chemistry, biology and material science. Noble metals, especially gold (Au) and silver (Ag) nanoparticles, exhibit unique and tunable plasmonic properties; the control over these metal nanostructures size and shape allows manipulating their LSPR and their response to the local environment. In this review, we will focus on Ag-based nanoparticles, a metal that has probably played the most important role in the development of the latest plasmonic applications, owing to its unique properties. We will first browse the methods for AgNPs synthesis allowing for controlled size, uniformity and shape. Ag-based biosensing is often performed with coated particles; therefore, in a second part, we will explore various coating strategies (organics, polymers, and inorganics) and their influence on coated-AgNPs properties. The third part will be devoted to the combination of gold and silver for plasmonic biosensing, in particular the use of mixed Ag and AuNPs, i.e., AgAu alloys or Ag-Au core@shell nanoparticles will be outlined. In the last part, selected examples of Ag and AgAu-based plasmonic biosensors will be presented.

Keywords: silver nanoparticles; synthesis; coating; alloy; core@shell; LSPR; biosensors

---

1. Introduction

The first use of silver (Ag) as an antimicrobial and antibacterial agent goes back to the ancient Greek and Roman Empire [1,2]. At that time, the medicinal and preservative properties of silver were mainly used to protect vessels from bacterial attacks and to make water and other liquids potable [1,3–5]. Globally, it was already known to be an efficient weapon against the growth of pathogen factors [6]. The antimicrobial effect of silver arises from the interaction of silver ions with thiol groups of vital bacterial enzymes and proteins that lead to cell death [4,5,7]. Over the past decades, silver has been engineered into nanoparticles (NPs) (at least one dimension is smaller than 100 nm) [8]. Although metallic NPs have been present in artefacts for a very long time [9,10], as attested by medieval stained glasses, and even earlier as for the Lycurgus Cup of the British Museum in London, dated from the 5th century, scientific knowledge about NPs is quite recent [11]. Synthesis of metallic NPs can be achieved according to two distinct nanofabrication methods. On the one hand, top-down approaches involve physical processes such as lithography or chemical processes controlled by external experimental parameters to create nanoscale structures starting from larger dimensions to the nanometer range [12]. This can be achieved by milling or high pressure homogenization [13]. On the other hand, the bottom-up approaches use atoms or small molecules as the building blocks of multi-level structures to build up more complex nanoscale assemblies or directed self-assemblies that perform various operations [14]. This method is extremely valuable since it is free of waste or unused materials [15]. This can be achieved by controlled precipitation (or crystallization) and evaporation from a precursor [13,16]. Generally, top-down techniques produce NPs that are mostly crystalline but high energy or pressure is required to achieve nanometer range comminution, which may also lead to contamination if a milling medium is used. In contrast, bottom-up processes involve dissolution, followed by precipitation or drying. The mechanical energy input is thus minimal, and the resulting NPs can be crystalline or amorphous, depending on the synthesis conditions. Metallic NPs feature unique physical properties such as high ratio-to-surface area and volume. Moreover, the confinement effect confers reactivity as well as mechanical, electromagnetic, chemical and optical properties that differ from those of the bulk metals [17–19]. Indeed, particles properties change drastically at the nanometer scale [11]. Metallic NPs find applications in various fields, from catalysis to the detection of biological molecules in solution [8,20]. In the biomedical field, they can be used in drug delivery, photothermal therapy, or imaging [21–25]. In what follows, we will focus the use of NPs for biosensing applications.

Among all metallic NPs, gold (Au) and silver (Ag) nanoparticles exhibit the most interesting physical properties for biosensing [26,27]. Even if gold nanoparticles (AuNPs) remain the most studied for this application area because of their good chemical stability and biocompatibility [20,28,29], silver nanoparticles (AgNPs) offer better results in terms of sensitivity [30]. One of the most characteristic physical properties of metallic NPs is the localized surface plasmon resonance (LSPR), which is responsible for the bright color of the nanoparticle colloidal suspensions [26,27,31–34]. Indeed, AuNPs and AgNPs are specifically investigated for their optical properties thanks to their strong interactions with light [35–40]. The electrons at the surface of the metallic NPs undergo a collective oscillation when irradiated at a specific wavelength, called surface plasmon resonance (SPR), resulting in the appearance of the electromagnetic field localized on the NPs [30,31,41]. When the oscillations of the electromagnetic field of an incident electromagnetic wave are in resonance with those of the local electromagnetic field of the NPs, the LSPR phenomenon occurs, which is characterized by the resonance oscillation frequencies. Thus, LSPR is the consequence of the confinement of the electric field within a small metallic sphere whose radius is much smaller than the wavelength [10]. This property can be tuned by controlling parameters such as shape, size, uniformity and surface coating [27,31] and is often used for biosensing applications in the field of biology, biomedicine and biochemistry [42]. For this purpose, AgNPs of different shapes and sizes, from the simplest to the most sophisticated, can be readily obtained thanks to the large range of techniques now available that will be presented later in this review for a conceptual opportunity of biosensing. Owing to their plasmonic properties, metallic nanoparticles are also responsible for enhancing Raman scattering of molecules adsorbed at their surface, giving rise to the so-called surface enhanced Raman spectroscopy (SERS) [32,43], a powerful vibrational spectroscopy with impressive enhancement factors of up to 14–15 orders of magnitude [44]. It is worthy to note that this phenomenon is very different from propagative SPR or surface plasmon polariton (SPP) that occurs at the plane surface of large metallic structures, or on metallic nanowires, on which one direction is regarded as infinite. Colloidal suspensions of small spherical AuNPs (10 nm diameter) are red-colored and display an absorption band at 520 nm, while similar AgNPs are yellow and absorb around 380 nm [45]. As these two absorption bands are in the visible part of the electromagnetic spectrum, they allow a colorimetric detection of biomolecules by inducing changes in the position (and possibly in the intensity) of the LSPR band.

As stated above, the position of the LSPR band of AuNPs and AgNPs depends on their size, uniformity, shape, dispersion, composition (ratio Au:Ag), and also on the dielectric constant of the surrounding medium [9,26,30,32,43,46,47]. Therefore, modifying any of these parameters induces wavelength shifts. For example, size increase shifts the LSPR band to higher wavelength, i.e., red-shifted [47,48]; NPs aggregation also induces a red-shift. While isotropic (spherical) NPs have a unique absorption band, anisotropic ones can display several absorption bands [47,48]. Two formats are typically encountered for colorimetric and plasmonic biosensors, i.e., aggregation-based assays [49–51] and LSPR-based ones [38]. The refractive index sensitivity (RIS), expressed in nm/refractive index unit (RIU) (nanometer per refractive index unit) is a measure of the shift in wavelength of the LSPR peak: the more the peak is shifted for small variations of refractive index (RI), the more sensitive the biosensor is (i.e., the highest the sensitivity is). AgNPs are described as more sensitive than AuNPs for the second biosensing strategy [52]. Indeed, a study showed that the RIS for AgNPs and AuNPs increased from 153 to 265 nm/RIU and 128 to 233 nm/RIU, respectively, for sizes 5 to 50 nm [53]. However, combining the two metals is very attractive and offers a wide range of possibilities.

In what follows we discuss multiple strategies to produce AgNPs of various sizes and shapes that are both key factors contributing to the modulation of major optical properties. Then, we will explore different coatings (organic, polymer, inorganic) on silver nanoparticles and highlight the influence of AgNPs coating process on their fate, stability, and toxicity in a given environment as well as their properties regarding plasmonic biosensing applications. We will also cover the use of mixed Ag and Au NPs (i.e., AgAu alloy and Ag-Au core@shell structures) for optical biosensing. Finally, we will present selected examples of Ag and AgAu-based plasmonic biosensors and highlight the merit of silver-containing nanoparticles in this area.

2. Engineering Silver Nanoparticles for Biosensing: Shape-Properties Correlation

Much less effort has been put on the use of AgNPs for biosensors compared to AuNPs. Indeed, 85,570 papers are referenced in the SciFinder™ database with the keyword "gold nanoparticles" compared to 63,770 for "silver nanoparticles". Combined with the keyword "plasmonic biosensors", these numbers are reduced to 424 and 112 papers, for AuNPs and AgNPs, respectively.

The synthesis of AgNPs is achieved through chemical, physical, or biological strategies. The review written by B. Khodashenas and H. R. Ghorbani in 2015, summarizes the wide range of synthetic methods reported to date as a function of the desired nanoparticle's shape [54]. We summarize in Table 1 the main strategies employed for AgNPs engineering and the related size, shape, and applications.

Table 1. Summary of the main strategies employed to synthesize silver nanoparticles (AgNPs) along with the corresponding shape and size, as well as their respective applications.

Shapes	Sizes (nm)	Synthesis Methods	Applications	Ref.
Spheres	10–100	Chemical reduction	Plasmonic and sensing; catalysis; antimicrobial	[55,56]
Triangles	Width: 100 Height: 50	Chemical reduction; nanosphere lithography (NSL)	Plasmonic and sensing; analytical devices (SERS); photovoltaics; molecular detection (Alzheimer disease)	[56–59]
Nanocubes	20–45	Polyol process	Cysteine sensing by plasmons of silver nanocubes; analytical devices (SERS)	[54,60,61]
Nanowires	Length: 60	Polyol process	Plasmonic and molecule sensing; provide conductive coatings (transparent conductors and flexible electronics)	[54,62,63]
Nanorods	Length: 250–300	Photochemical; thermal; oxidation reduction growth (ORG)	Plasmonic and sensing; analytical devices (SERS)	[54,64]
Nanobars	Length: 100	Polyol process	Plasmonic and sensing; analytical devices (SERS)	[54,56]
Pyramides	Edge length: 50–200	Chemical reduction	Plasmonic and sensing; analytical devices (SERS)	[54]
Flower-like	200–300	Wet-chemical method	Analytical devices (SERS); catalysis	[54,65,66]

2.1. AgNPs Synthesis by Chemical Reduction Using Citrate and/or Ascorbate

Nowadays, among the wide range of synthetic methods, the chemical reduction by the bottom-up approach is the most common method to prepare AgNPs. The reaction is performed in either an aqueous or an organic solvent. The commonly used method is inspired from the Turkevich synthesis for AuNPs [67], and was first introduced by Lee's group in 1982 [68]. It involves a silver precursor, usually an inorganic compound such as silver nitrate ($AgNO_3$) that liberates silver ions (Ag^+) in solution, then, trisodium citrate ($Na_3C_6H_5O_7$) reduces silver and further stabilizes the resulting AgNPs [69]. However, this method often leads to polydisperse nanoparticles and several works attempted to reduce the dispersity [69,70]. Indeed, many factors have been demonstrated to play an important role on the size, shape and color of NPs such as temperature and acidity of the solution [71]. Actually, the acidity of the solution has a strong influence on the AgNPs' shape. In 2009, Dong et al. showed that the shape of AgNPs was significantly influenced by the pH using citrate as a reducing as well as stabilizing agent (Figure 1A) [72]. They found out that the shape of AgNPs at a high pH was a mixture of spherical and rod-like particles while the predominant shapes were triangular and polygonal particles at low pH (Figure 1B) [71,72].

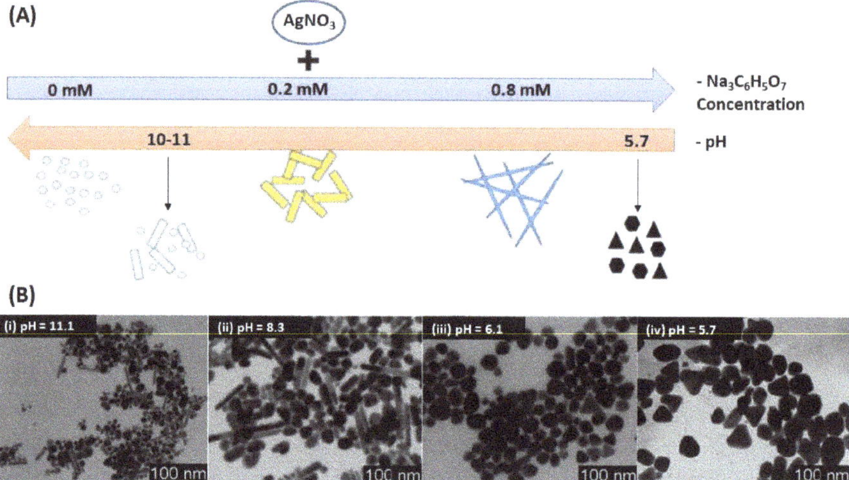

Figure 1. AgNPs synthesis using citrate: (**A**) Experimental conditions affecting the silver nanoparticle AgNPs shape and (**B**) TEM images of the AgNPs synthesized at different pH values: (i) 11.1, (ii) 8.3, (iii) 6.1 and (iv) 5.7. Adapted from [72]. Copyright (2009), American Chemical Society.

This observation was confirmed by Qin et al. in 2010 as more spherical particles were obtained when the pH was increased by using ascorbate as a reductant and citrate as a stabilizer [73]. Furthermore, they demonstrated that NP size varied as a function of pH. Indeed, AgNPs were smaller due to the increased reducing activity of citrate or ascorbic acid (AA) when the pH was increased that in turn decreased the number of nuclei [71,73]. These studies also showed that the type of reductant affects the shape of the AgNPs due to the pH-dependent redox potential and thus the pH-dependent reduction rate of the precursor (Ag^+). A more precise adjustment of the equilibrium between nucleation and growth allows better control over the shape of AgNPs. The use of ascorbic acid as reducing agent tends to afford spherical particles over the pH range between 6 and 11 conversely to citrate (Figure 2).

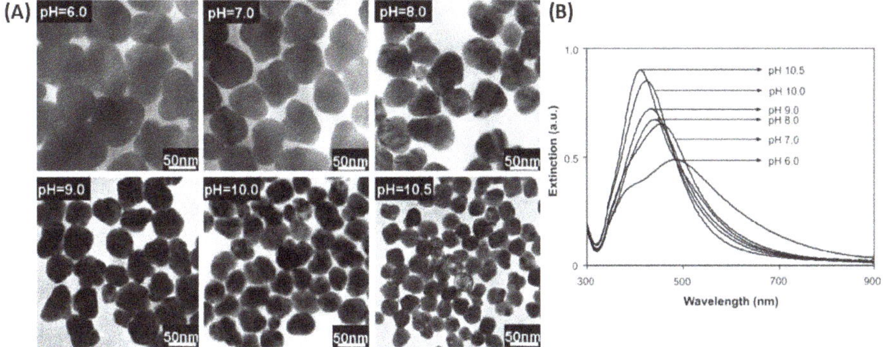

Figure 2. (**A**) TEM images and (**B**) UV-vis spectra of the AgNPs prepared at pH 6.0, 7.0, 8.0, 9.0, 10.0 and 10.5 by using ascorbate as reductant. Adapted from [73]. Copyright (2010), Elsevier B.V. All rights reserved.

Another study showed that the position of the LSPR band shifted as a function of the acidity [71]. Indeed, the LSPR band intensity increased and the band blue-shifted as the pH increased. The bands became sharper, and the nanoparticles size decreased accordingly as demonstrated by the TEM images in Figure 2. Finally, citrate was also used as reducing agent to form Ag nanoshell (AgNS) on a silica sphere [74–76]. The thickness of the silver shell could be tailored by varying the number of deposition cycles. Such a weak reducing agent was a prerequisite to the growth of a silver shell, so that the Ag seeds grew only in size during the Ag reduction whereas no new nucleation centers were introduced, which ensured the minimal amount of Ag colloids in the suspension accompanying the AgNS growth.

2.2. AgNPs Synthesis: Anisotropic Shapes

Investigations on anisotropic shapes and morphologies of NPs have increased during the last decade, most often relying on the development of seed-mediated synthetic methods [77]. Such syntheses of anisotropic NPs have attracted interest because the structural, optical, electronic, magnetic, and catalytic properties are different from, and most often superior to, those of spherical NPs for a comparable size [78]. In particular, the most striking properties of anisotropic and hollow NPs lie in the appearance of a plasmon band at a longer wavelength (near-infrared region) than that of spherical NPs [79,80]. Different shapes and sizes of AgNPs can be synthesized thanks to the large range of techniques now available [72,77,81–85]. Inspired by gold nanorods synthesis, silver nanorods (AgNRs) were synthesized following a process involving the reduction of $AgNO_3$ with sodium borohydride ($NaBH_4$) in the presence of citrate followed by growth of seeds into NRs in the presence of AA and cetyltrimethylammonium bromide (CTAB) [86]. In 2011, Zaheer and Rafiuddin achieved the synthesis of flower-like silver NPs at room temperature by a wet chemical reduction strategy [87]. It involves the use of AA as reducing agent and CTAB. Such a shape is the result of the aggregation of small NPls and NRs. Flower-like AgNPs were used as SERS substrates and showed high sensitivity to rhodamine 6G.

Several other studies showed that it is possible to make the transition from spherical to nanowire particles through nanorods by modulating the experimental conditions such as temperature, Ag precursor concentration, pH of the solution, reducing agent (citric acid, AA, and $NaBH_4$), and the method (chemical, physical or biological) (Figure 3) [54,56,88].

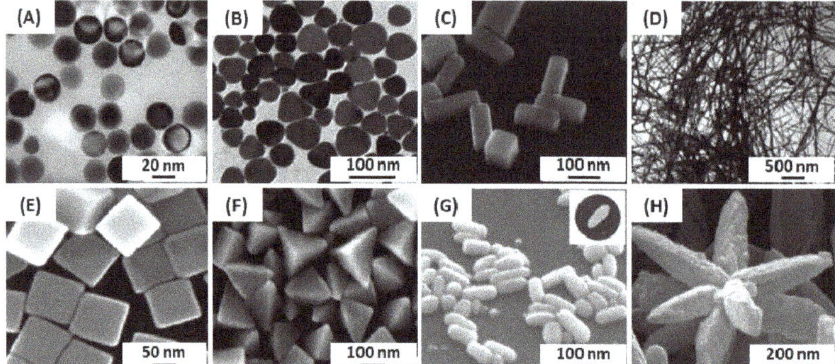

Figure 3. TEM images of silver nanoparticles with different shapes: (**A**) nanospheres, (**B**) nanoprisms, (**C**) nanobars and (**D**) nanowires. SEM images of (**E**) nanocubes, (**F**) pyramids, (**G**) nanorice and (**H**) nanoflowers. Adapted from [54,56]. Copyright (2009), Springer Science Business Media, LLC.

A particularly interesting morphology for the development of LSPR biosensors is that of silver nanoplates (AgNPls), in which the lateral dimensions are larger than their height, so that they possess an extremely high degree of anisotropy (Figure 4A). Although the systems used for LSPR biosensing have been mainly ordered arrays of triangular NPls (TNPls) prepared by nanosphere lithography (NSL), wet chemistry techniques are one of the most widely used methods with a tight control of size and shape [30,89,90]. The wet chemistry approaches to synthesize AgNPls are light-mediated methods that relate to the use of visible light as a driving force for the oriented attachment of preformed nanoparticles. Some chemical reduction methods are based on the reduction of Ag^+ on Ag seeds using a weakly reducing agent (citrate or ascorbate), in the presence of CTAB, by analogy to the well-known growth of Au nanorods [90,91]. AgNPls optical resonance can be tuned from 500 to 1100 nm by precisely controlling the plate diameter and thickness [92]. AgNPls have extremely large absorbing and scattering cross-sections across the visible and near-infrared (NIR) regions (Figure 4B) [93]; they have applications in SERS, photovoltaics, molecular detection, and photothermal-based therapies [30,89,94].

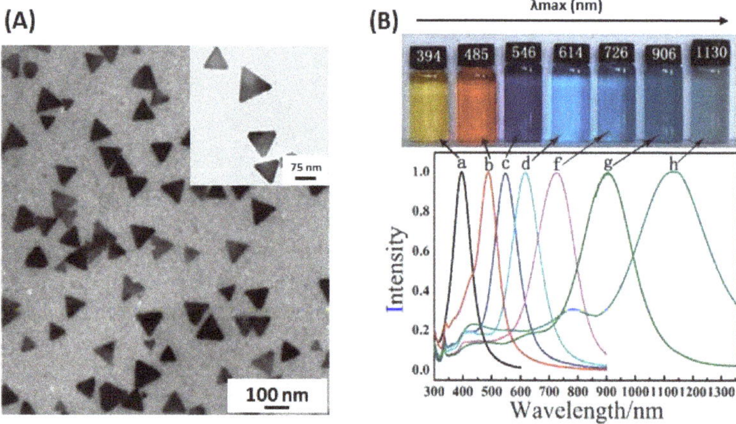

Figure 4. (**A**) TEM image of AgNPls. (**B**) Dispersions of Ag (a) sphere and (b–h) nanoplate colloids with different colors and corresponding UV-vis-NIR extinction spectra that reflect the ability to tune the plasmon resonance of the nanoplates across the visible near-infrared (NIR) portion of the spectrum (500–1100 nm). The nanoplate optical resonance and size are tuned according to different rounds of Ag seed growth. Adapted from [90,93]. Copyright (2008, 2013), Royal Society of Chemistry.

Nanoprisms (NPrs) seem to present a classic triangular shape, but a closer observation showed that the triangles apexes were flat. Compared to spherical nanoparticles, NPrs with flat apexes and (111) crystal planes, exhibited greater antibacterial property [95]. Ag triangular nanoparticles may be also fabricated by NSL. In fact, Haes and Duyne demonstrated a very good RIS by tuning the shape and the size, and a short-range, sensing length scale determined by the characteristic decay length of the local electromagnetic field [57]. NSL is widely used to get mono-disperse, surface-confined Ag triangular NPs. It is based on the creation of a single layer crystal nanosphere mask with a suspension containing monodisperse spherical colloids (polystyrene) onto a glass substrate [96,97]. Then, a drying step is required, and the mask is formed. After that, a film of silver material is deposited over the support and the mask is then removed by a step of sonication in an adequate solvent, as shown in Figure 5. The size of the nanotriangles is controlled by the diameter of the nanospheres deposited [97].

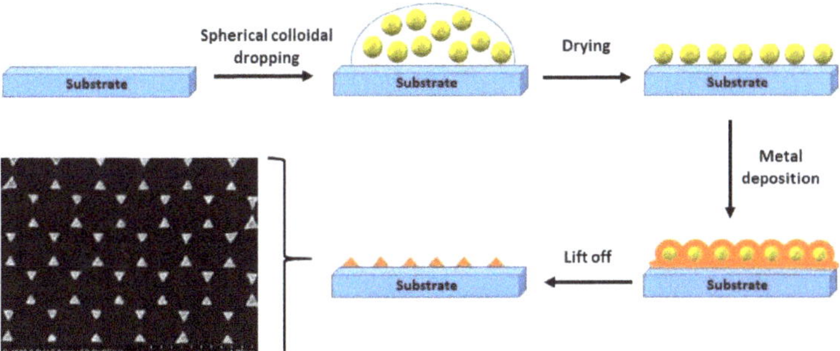

Figure 5. Diagram of the nanosphere lithography (NSL) mechanism. SEM image of topography of the triangular Ag nanoparticles fabricated by NSL. Adapted from [96]. Copyright (2011), publisher and licensee Dove Medical Press Ltd.

2.3. AgNPs Synthesis: Chemical Reduction Using Unconventional Ligands

In 2002, it was reported that silver nanocubes were synthesized with a polyol process (Figure 6A), which involves the reduction of silver thanks to the hydroxyl groups of ethylene glycol [82]. The alcohol acts both as solvent and reducing agent [54]. A capping agent is then added; generally polyvinyl pyrrolidone (PVP) whose role is to build the cubic shape of the NPs [98–100]. Later, Tao et al. (2006) found that the ratio between Ag^+ and the number of repeating units of PVP defined the geometry of the NPs [101]. The nanocube formation was favorable when the ratio was high. On the contrary, the nanowire geometry would have been favorable. In the same year, Siekkinen et al. found out that adding a small quantity of sodium sulfide (Na_2S) or sodium hydrosulfide (NaHS) speeds up the reaction, from 16–26 min to 3–8 min (Figure 6B,C) [102].

AgNRs were synthesized by using a method called oxidation reduction growth (ORG) [64]. Firstly, a thin silver film is deposited on a silica surface with a relatively constant flow of argon gas. Then, silver oxide seeds are formed during sputtering. In the sputtering process, the temperature increased to reach 200 to 300 °C. Hence, silver oxide dissolved and released oxygen. It allowed silver nanorods to grow without oxygen (Figure 7).

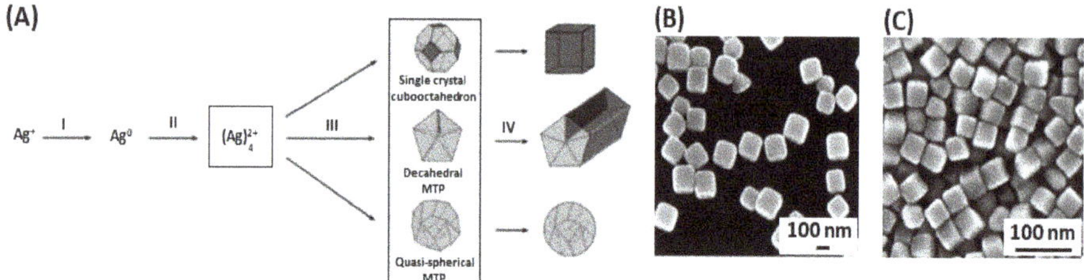

Figure 6. (**A**) illustration of (I) Ag$^+$ reduction by polyol process; (II) formation of Ag clusters; (III) seed nucleation; and (IV) seed growth into nanocubes, nanorods or nanowires, and nanospheres. SEM images of Ag nanocubes synthesized by mixing AgNO$_3$ and polyvinyl pyrrolidone (PVP) *via* polyol process: (**B**) without and (**C**) sulfide-assisted synthesis (reaction time: 45 min vs. 7 min, respectively). Adapted from [98,102]. Copyright (2004), © WILEY-VCH Verlag GmbH & Co. KGaA, Weinheim. Copyright (2006), Elsevier B.V. All rights reserved.

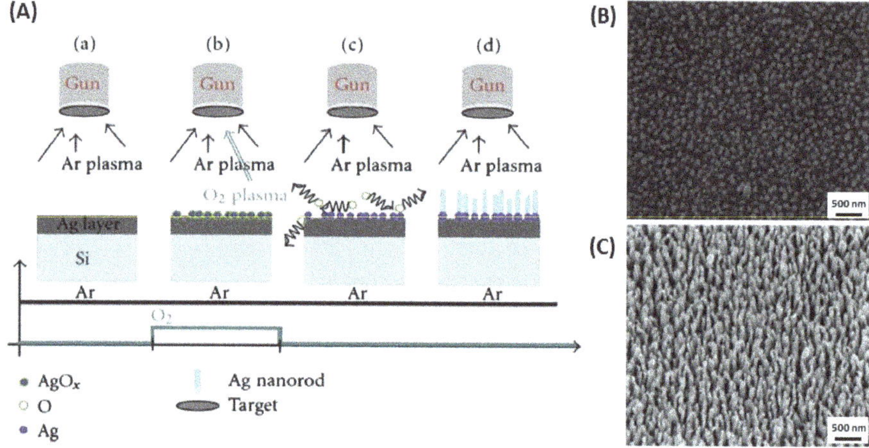

Figure 7. (**A**) Synthesis of silver nanorods by sputtering process: oxidation reduction growth (ORG). SEM images of (**B**) Ag nuclei and (**C**) AgNRs arrays. Adapted from [64].

3. Coating of Silver Nanoparticles

There are very few studies on bare AgNPs as plasmonic biosensors. One of the reasons concerns their toxicity, even if most biosensors operate ex vivo. AgNPs toxicity was extensively described in a book published in 2019 [103]. The second reason, and most probably the major limitation for bare AgNPs use in biosensing, is their poor stability and less straightforward surface chemistry [104,105]. To overcome these limitations, AgNPs were coated by a large variety of compounds; the coating process has a marked influence on the fate, stability, and toxicity of AgNPs in a given environment [106–108]. The coating of the NPs provides electrostatic, steric, or electrosteric repulsive forces between particles, which allows them to resist aggregation phenomena [105]. In the literature, various coating methods have been explored to cover AgNPs with an organic or an inorganic shell and highlighted the interest of coating AgNPs for plasmonic biosensing applications (Table 2). Hence, both the nature of the coating reactant and the thickness of the coating layer have a decisive influence on the optical properties of the NPs. In what follows, we will present examples of AgNPs coatings and discuss their influence regarding plasmonic biosensing. Interest will be first brought to organic-coated (excluding polymers) AgNPs, then to polymer-coated ones in order to improve electrostatic, steric and electrosteric stabilization of AgNPs. Finally, a brief overview of silica coating on NPs will be made.

Table 2. Correlation between the coating nature and AgNPs stabilization mode.

Shape	Type of Coating	Size (nm)	AgNPs Stabilization	Ref.
Spherical	Citrate	14–20	Electrostatic	[105]
Triangular	Citrate	10–20	Electrostatic	[98]
Spherical	Plant root extract	30–55	Electrosteric	[109]
Spherical	Sodium dodecyl sulfate (SDS)	26	Electrosteric	[110]
Spherical	Tween 80	17–42	Steric	[110]
Nanocube	PVP	80	Steric	[111]
Nanobeam	PVP	17–70	Steric	[111]
Triangular	Chitosan	115–123	Steric	[91]
Spherical	PVA	8–46	Steric	[55]
Spherical	Silica	55–65	Electrostatic	[112]
Triangular	Silica	40–50	Electrostatic	[113]

3.1. Organic Coatings

AgNPs synthesis typically uses organic compounds to promote stabilization and prevent aggregation of the particles by adsorption or covalent attachment to the particles surface. In the literature, they are often referred to as "capping agents" when they are applied during synthesis. It was proven that they have an effect on the size and shape control of the AgNPs [98]. Therefore, the function of organic coating in the stabilization and the growth of AgNPs is clearly essential for their further properties [107]. There are different possible shapes of AgNPs including quasi-spheres, nanotubes, rods, or triangular nanoplates (TNPls) which also means different coating methods with capping agents of various chemical nature (Figure 8) [114].

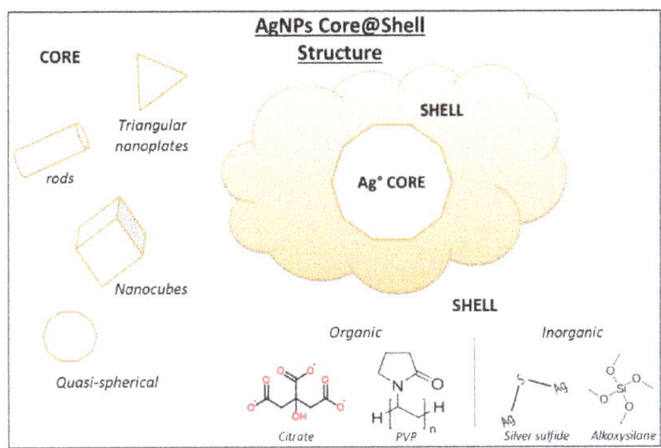

Figure 8. Silver nanoparticles core@shell structure. Adapted from [114]. Copyright (2012), American Chemical Society.

Natural organic matter (NOM) is a quite interesting example of organic coating of AgNPs [115,116]; NOM significantly influenced the stability and the surface properties of NPs, and had in turn a direct effect on the transport and the AgNPs toxicity in aqueous systems. In 2015, Gunsolus and his co-workers used NOM to stabilize citrate- and PVP-capped AgNPs against aggregation [117]. AgNPs incubated with NOM showed higher primary extinction peak intensity, which means a larger population of monodisperse particles, and slower aggregate formation by observing the secondary extinction peak (Figure 9). However, we could find no example of use of NOM-coated AgNPs as LSPR biosensors, possibly because the NOM shell is too large (up to 150 nm in Reference [117]) and, as LSPR is a short distance effect, the molecular phenomena occurring at the NOM shell no longer affect the LSPR signal. Besides, another green method was used to synthesize organic-coated AgNPs from extracts of soap-root plant as stabilizers and manna of hedysarum plant as reducer [118].

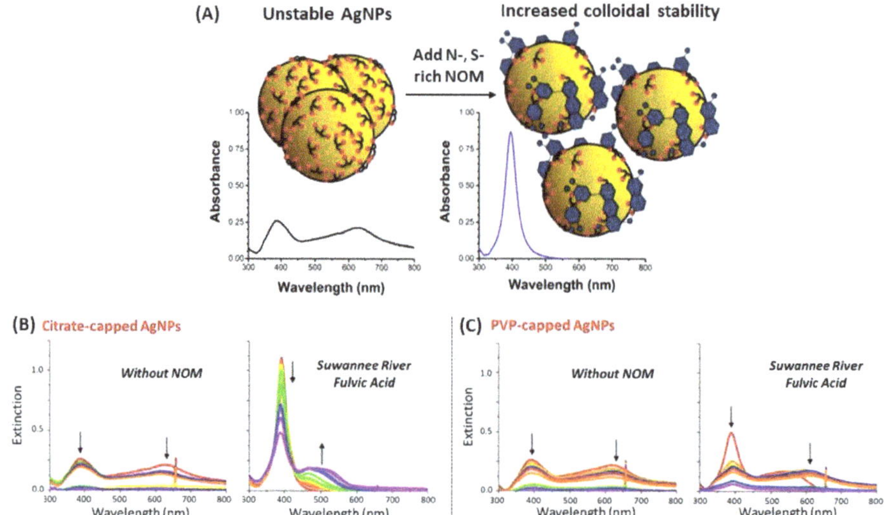

Figure 9. (**A**) Illustration of natural organic matter (NOM) interactions with the surface of silver nanoparticles according to the NOM's chemical composition and the affinity of the capping agent for the AgNP surface. Colloidal stability of (**B**) citrate- and (**C**) PVP-capped AgNPs in the absence or the presence of NOM from various origins. Adapted from [117]. Copyright (2015), American Chemical Society.

Many studies investigated the use of thiol-capping agents as anchoring groups for stabilizing and protecting AgNPs [108,119,120]. The thiol-capping agents are grafted to the AgNPs surface through Ag-S chemical bonds to form the external layer suggesting a core@shell morphology with an Ag core surrounded by Ag_2S-like shell. A study showed that stabilized AgNPs, by the organic thiol, allylmercaptane (AM), were synthesized with different Ag/S molar ratios in the presence of tetraoctylammonium bromide (TOAB) and $NaBH_4$ [119] *via* modified Brust–Schriffin method [121]. It has been shown that the increase in Ag/AM ratio led to an increase of the Ag_2S layer thickness, and thus larger AgNPs were obtained, while the external AM layer remained unchanged (Figure 10A) [119]. Desireddy et al. prepared ultrastable AgNPs with a uniform size from the reduction of soluble precursor, which uses a protecting shell of p-mercaptobenzoic acid in semi-aqueous solution in the presence of $NaBH_4$ and a coordinating solvent. This approach showed better results regarding the stability, purity and yield in very large quantities compared to those for AuNPs, due to an efficient stabilization mechanism [108]. Another approach was used by Cheng et al. using thiol-modified metal-organic framework (MOF) [120]. Herein, MOF was used as a host matrix to obtain AgNPs by using the stabilization ability of the thiol group to prevent further aggregation (Figure 10B). By controlling the initial loading amount of silver ions in the cages of thiol-MOF, monodispersed AgNPs were encapsulated in frameworks by reducing Ag^+ with $NaBH_4$, while adjusting sizes of particles from 5.3 nm to 3.9 nm, which is difficult to achieve for AgNPs because of their strong tendency to aggregate.

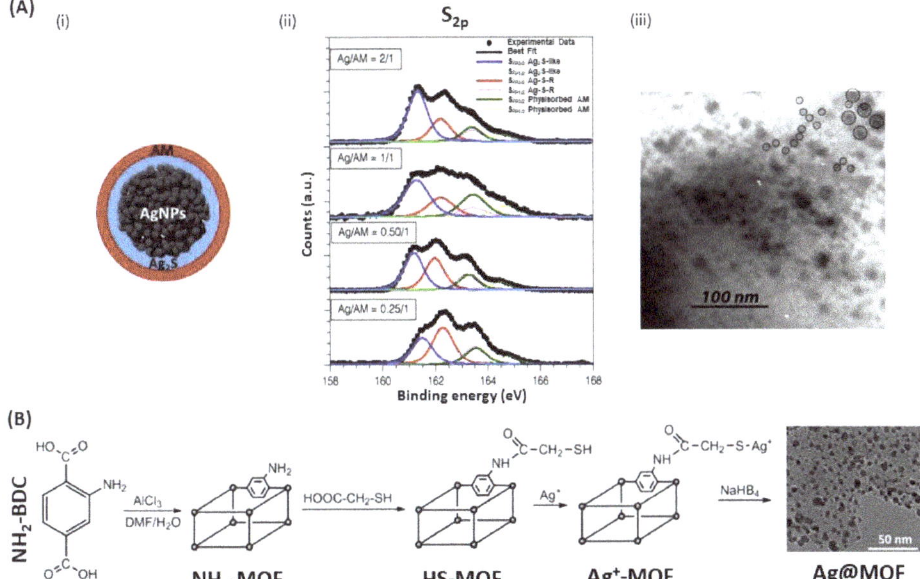

Figure 10. (**A**) Allylmercaptane-stabilized AgNPs: (i) core@shell morphology for allylmercaptane-(AM)-functionalized AgNPs through Ag-S chemical bonds to form the external layer, (ii) XPS spectra of Ag@AM with four different Ag/thiol ratios, and (iii) TEM images of AgNPs with Ag/AM molar ratio equal to 2/1 (AgNPs dimensions are 9 ± 3 nm and a population of NPs aggregated of 18 ± 6 nm. (**B**) Illustration of the encapsulation of AgNPs in thiol-modified metal-organic framework (MOF) as a host matrix. Adapted from [119,120]. Copyright (2012), American Chemical Society. Copyright (2015), Royal Society of Chemistry.

3.2. Polymer Coatings

Polymers are molecules that can adopt various conformations in solution. The chain swelling can be modulated by the temperature [122]. This aspect of polymers properties has been investigated because the main interest of polymer coating comes from steric interactions. Indeed, polymers, either grafted or adsorbed on NPs, promote dispersity in the NPs solution [123]. It has long been established that polymers with a large molecular weight and a high grafting density tend to increase the colloidal stability [124,125].

Poly(ethylene) glycol (PEG) is one of the most studied polymers as stabilizing or coating agent for NPs [123,126,127]. This neutral, hydrophilic and biocompatible polymer has been approved by the Food and Drug Administration (FDA) for biomedical and pharmaceutical applications [128]. PEG improves the AgNPs dispersity in physiological conditions by steric hindrance and prevents nanoparticles aggregation [107,115,129]. Figure 11 represents only one of many ways of PEG coating by a green method [130]. Colloidal stabilization for PEG-coated AgNPs probably occurs thanks to the presence of VdW interactions:

$$Ag^+_{(aq)} + PEG_{(aq)} \rightarrow [Ag(PEG)]^+_{(aq)}$$

$$2\,[Ag(PEG)]^+_{(aq)} + CH_2OH(CHOH)_4CHO \rightarrow 2\,[Ag(PEG)]_{(s)} + CH_2OH(CHOH)_4COOH$$

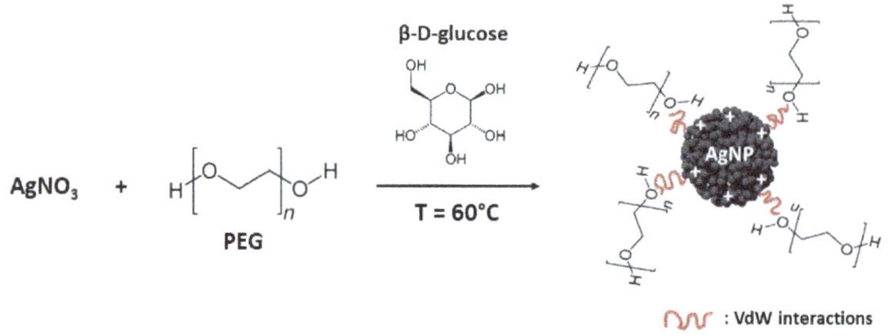

Figure 11. Poly(ethylene) glycol (PEG) coated method of silver nanoparticles. Inspired from [130].

Another polymer, called chitosan, is widely used to coat NPs because of its good biocompatibility [131,132]. It shows a good affinity for the Ag surface and confers a high stability and dispersibility to the AgNPs [91,133]. It also shows shape-directing properties by influencing the shape of the particles from spherical to triangular.

3.3. Silica Coating

Among all coating materials used for plasmonic nanoparticles capping, silica occupies a pro-eminent position for multiple reasons [134–136]; first, silica provides a biocompatible protective shell, tunable in thickness, preventing aggregation due to electrostatic repulsion and stable in numerous solvents; second, silica synthesis is largely mastered especially through sol-gel and/or Stöber process to achieve a nanometric control of the thickness, the porosity, and the homogeneity; lastly, the presence of silanol groups on silica surface simplify the further chemical modification to introduce various surface functionalities (e.g., COOH, CHO, NH_2, or NCO) with readily available coupling agents [137,138]. The Mie theory already predicts effects of silica shell thickness on NPs optical properties in various solvents [139]. There are several ways to coat NPs with silica, among which the modified Stöber process that enables to control the shell growth over a short time period (Figure 12) [135,140,141].

Figure 12. Modified Stöber method for coating of AgNPs with silica.

Kobayashi, Liz-Marzán and their co-workers synthesized SiO_2-coated AgNPs by sol-gel reaction of tetraethyl orthosilicate (TEOS) [140]. They observed that the shell thickness was controlled through TEOS concentration and observed an increasing red-shift of the LSPR band for thicknesses in the range 28 to 48 nm. Larger silica shell thicknesses, 57–76 nm, induced a blue-shift of the plasmon band as well as a decrease of its intensity, which means that larger silica shells promote significant scattering at shorter wavelengths. Their findings were consistent with the theoretical spectra calculated by the Mie theory. Coating of anisotropic AgNPs, e.g., triangular nanoplates (AgTNPls) is more challenging as methods for silica coating of spherical AuNPs were found to be unsuitable for triangular nanoplates [113]. Silica coating of AgTNPls was achieved through a modified Stöber approach using TEOS as the alkoxide precursor and various primers: diaminopropane priming followed by reaction with TEOS (Figure 13A) that allowed tuning the thickness of the silica shell in the range 7 to 20 nm, or mercaptopropyltrialkoxysilane (either ethoxy or methoxy, MPTES or MPTMS, respectively) priming followed by silica deposition from sodium silicate (Figure 13B). This latter method using MPTES conveyed the highest stability towards salt, while retaining RI sensitivity comparable to that of the original uncoated particles (Figure 13C).

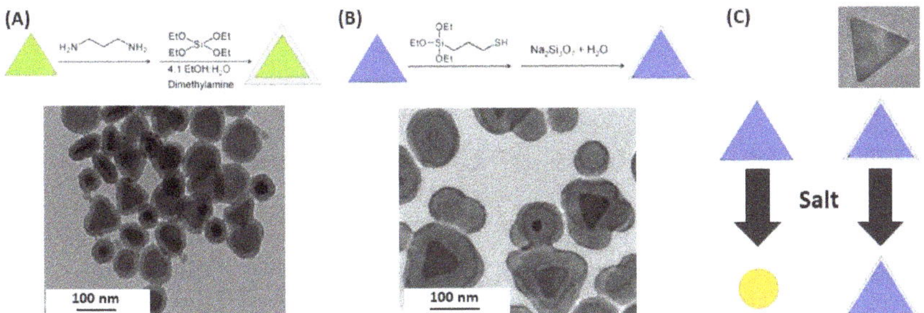

Figure 13. Silica coating of Ag triangular nanoplates by (**A**) diaminopropane priming and (**B**) (3-Mercaptopropyl)triethoxysilane (MPTES) priming followed by deposition from $Na_2Si_3O_7$ solution. (**C**) The silica shell using MPTES on triangular nanoplates (AgTNPls) allows withstanding salts without adversely affecting refractive index (RI) sensitivity in relation to original uncoated particles. Adapted from [113]. Copyright (2013), Elsevier Inc. All rights reserved.

The major interest of these particles is the wide tunability of the plasmonic energies which could have great attention in the development of biosensors [134]. The interest of silica coating was highlighted by Pratsinis and his group [142], who confirmed that silica coating prevented AgNPs agglomeration or flocculation then investigated of the plasmonic Ag@SiO$_2$ NPs the toxicity against a model biological system (*Escherichia coli*) and concluded that it was blocked by coating nanosilver with a silica shell about 2 nm thick. The method used for silica coating is different from those described above as they used a flame aerosol method using hexamethyldisiloxane as silica precursor (Figure 14). To predict the LSPR biosensing performances, they measured the lambda shift upon the adsorption of bovine serum albumin (BSA). The response was improved but this improvement seems to arise from a better dispersion and therefore a higher amount of protein adsorbed, even if no experimental data was provided to confirm this hypothesis [142].

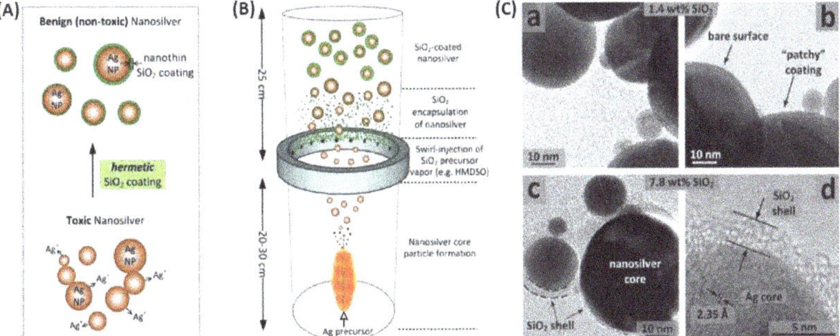

Figure 14. (**A**) Effect of a hermetic SiO$_2$ coating on the flocculation and toxicity of nanosilver particles. (**B**) Illustration of the nanosilver encapsulation with a hermetic SiO$_2$ coating using hexamethyldisiloxane as silica precursor in a flame aerosol reactor. (**C**) TEM images of the (a,b) 1.4 wt.% and (c,d) 7.8 wt.% SiO$_2$-coated nanosilver. Adapted from [142]. Copyright (2010), WILEY-VCH Verlag GmbH & Co. KGaA, Weinheim.

To conclude this part, the benefit from silica coating is well-established, in terms of chemical and colloidal stability and reduced toxicity. The protective silica shell has a limited effect on biosensing ability as long as the thickness of the layer is limited to few nanometers. Some aspects of silica coating, for instance the porosity, were not discussed herein, but are mentioned in the relevant review cited above. Beside these inputs from silica shells, there are no amplifying or synergetic effects in the plasmonic response of the Ag@SiO$_2$ NPs, conversely to coating or mixing with gold, discussed in what follows.

4. Plasmonic Nanoparticles Based on Silver and Gold: Alloy, Core@Shell, Nanocages and Nanoshells

Compared to the two other plasmonic nanometals, i.e., gold and copper, silver nanoparticles have a higher theoretical refractive index sensitivity (RIS) (Figure 15) [53,143]. It has been shown that RIS increased from 153 to 265 nm/refractive index unit (RIU), 128 to 233 nm/RIU, and 117 to 212 nm/RIU, respectively, for AgNPs, AuNPs, and CuNPs with sizes from 5 to 50 nm. Spherical AgNPs exhibit a stronger LSPR absorption with a peak at 400 nm, which is from five to 10 times more intense than the gold one at 520 nm [53,143]. Despite this better sensitivity, observed both experimentally and theoretically, AgNPs have several drawbacks for biosensing; in addition to their poor stability and biocompatibility, they display less sensing reversibility due to light alteration, which makes their use in repeated cycles less reliable than that of AuNPs [80].

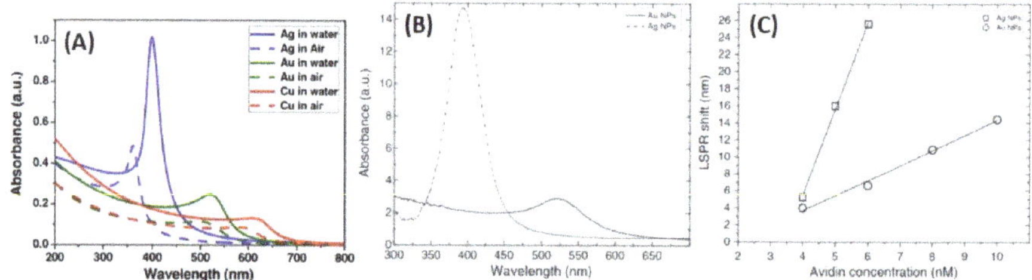

Figure 15. (**A**) Calculated extinction spectra of Ag, Au and Cu spherical NPs (20 nm) in different media. (**B**) Experimental extinction spectra of Ag and Au spherical NPs (10 nm) and (**C**) experimental response expressed as LSPR band shift of biocytin-coated Ag and Au spherical NPs (10 nm) in the presence of avidin. Adapted from [53,143]. Copyright (2014), Springer Science Business Media New York. Copyright (2012), Elsevier B.V. All rights reserved.

Many improvements have come about when AgNPs were combined to other metals and particularly to gold. Of course, the benefit from the previously discussed coating was effective with gold, but in addition, a synergy between these two plasmonic metals allowed for a better efficiency. This combination was mainly done by forming AgAu alloys or Ag@Au core@shell structures (Ag@AuNPs or Au@AgNPs). The main synthesis techniques and the resulting shapes and sizes are summarized in Table 3. In what follows, we will successively discuss AgAu alloys and Ag-Au core@shell structures synthesis. We will also cover the use of AgNPs as sacrificial templates to build gold nanocages (AuNC) or gold nanoshells (AuNS) for improvements of gold plasmonic biosensors.

Table 3. Summary of the main Ag-Au bimetallic NPs structures: synthesis techniques, size and shape.

Alloy/Core@shell	Shape	Synthesis Technique	Size (nm)	Ref.
AgAu alloy	Spherical	Chemical co-reduction of HAuCl$_4$ and AgNO$_3$ with sodium citrate	10–25 (0.27 < %Au < 1.00)	[144]
AgAu alloy	Spherical	Simultaneous laser ablation of Ag and Au colloids	5–50	[145]
AgAu alloy	Spherical	Metal evaporation on glass support and annealing (500 °C)	20–50	[146]
AgAu alloy	Spherical	UV laser radiation (193 nm) on silicate glass	5–40	[147]
AgAu alloy	Spherical core/alloy/shell	Sodium citrate reduction of Ag$^+$ on AuNPs and hydrothermal treatment	16–26 (T = 120 °C) 16–23 (T = 160 °C)	[148]
AgAu alloy	Elliptical (quasi-spherical)	Metal evaporation on glass support and annealing (350 °C)	Vertical radius: 4–12 Horizontal radius: 6–15	[149]
AgAu alloy	Nanoprisms (NPrs)	Nanosphere lithography and film deposition by thermal evaporation on a glass	150 (length) 50 (height)	[150]
Ag@Au core@shell	Spherical	Laser ablation of Au in a suspension of Ag colloids	30 (Ag core) 0.5–4 (Au shell)	[53]
Ag@Au core@shell	Nanoplates (NPls)	Electrodeposition of Au shell on Ag nanoplates (AgNPls)	50 (Ag core) 0.5 (Au shell)	[151]
Ag@Au core@shell	Hemispherical NPls	Cycles of electrodeposition of Au shell on AgNPls supported on ITO glass	100 (Ag core: width) 40 (Ag core: height) 1 (Au shell–20 cycles)	[152]
Ag@Au core@shell	Triangular nanoprisms (TNPrs)	Chemical reduction of HAuCl$_4$ by AA with PVP on silver TNPrs (AgTNPrs) by slow addition of HAuCl$_4$ solution	60 (Ag core) 1 (Au shell)	[153]
Au@Ag core@shell	Spherical	Deposition of Ag (chemical reduction) on AuNPs	10–15 (Au core) 1–10 (Ag shell) 30 (Au core)1–9 (Ag shell)	[40]
Au@Ag core@shell	Nanorods (NRs)	Sodium citrate and AA reduction of AgNO$_3$ on AuNRs	35 (Au core: length) 10 (Au core: width) 1–6 (Ag shell)	[154]
Au@Ag core@shell	NRs	Chemical reduction of AgNO$_3$ with AA on seed-mediated grown in NaBH$_4$ on AuNRs	60 (Au core: length) 30 (Au core: width) 1–3 (Ag shell)	[155]
Au@Ag core@shell	NRs	Chemical reduction of AgNO$_3$ with AA on seed-mediated grown in NaBH$_4$ on AuNRs	60 (Au core: length) 20 (Au core: width) 4 (Ag shell)	[156]
Au@Ag(@Au) core@shell	TNPrs	Sodium citrate and AA reduction of AgNO$_3$ on seed-mediated grown AuNPs supported on an ITO glass (followed by electrodeposition of a thin Au layer)	Initial Au@Ag TNPrs 30 (height) Au shell very thin when it is present	[157]

4.1. Silver-Gold Alloy Nanoparticles

Silver-gold alloy nanoparticles (AgAuNPs) are defined as a mixture of Ag and Au atoms, with no spatial distinction between the gold and silver parts. The chemical synthesis methods mainly consist in the co-reduction of AgNO$_3$ and HAuCl$_4$ with sodium citrate, which give spherical AgAuNPs [144,158]. The mole fraction of each metal in the alloy depends on the concentration of AgNO$_3$ and HAuCl$_4$ introduced in solution. In these conditions, small AgAuNPs (roughly 20 nm) can be synthesized [144]. Besides, simultaneous laser ablation of Ag and Au in colloidal suspension allows the synthesis of AgAuNPs, in the same range of size (Figure 16A) [145]. Another study demonstrated the AgAu alloy interdiffusion at the NPs interface, resulting in an intermediate alloy shell [148]. Indeed, a hydrothermal treatment is necessary during Ag$^+$ reduction at the surface of AuNPs for Ag diffusion in Au in order to obtain AgAuNPs. This phenomenon is dependent on the temperature, such as the growth of the Ag shell layer until the final structure: core/alloy/shell.

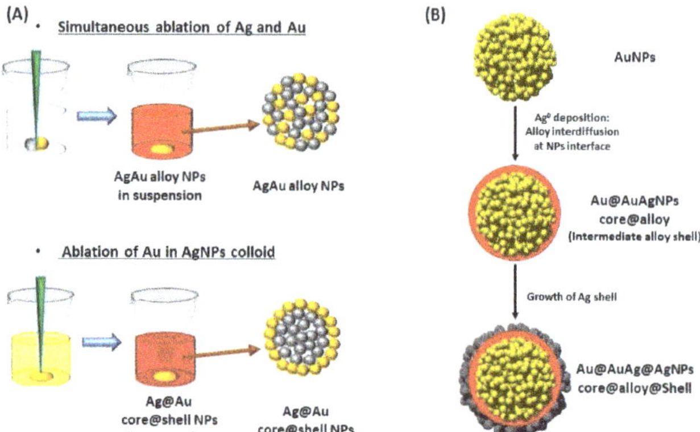

Figure 16. (**A**) Pulsed laser ablation in liquid: simultaneous ablation of Ag and Au to synthesize AgAuNPs and ablation of Au in AgNPs colloid to form core@shell structure (Au@AgNPs). Reproduced with permission from [145]. Copyright (2014,) Springer-Verlag Berlin Heidelberg. (**B**) Schematic illustration to form an intermediate AgAu alloy shell by interdiffusion at the NPs interface during hydrothermal treatment. Adapted from [148]. Copyright (2011), American Chemical Society.

AgAuNPs are also prepared by physical techniques, such as metal evaporation (electron beam), followed by thermal treatment which affords supported AgAuNPs on glass support [146,149]. Spherical or quasi-spherical AgAuNPs are obtained due to evaporation of Ag and Au layers in a vacuum chamber. Then, Au and Ag metallic atoms can be deposited on the glass substrate, following an island formation of AgAuNPs, because of the stronger interactions between Ag and Au atoms, than with the glass substrate [149]. The annealing post-treatment increases the crystallinity of AgAuNPs, but destroys completely pure AgNPs, previously synthesized in the same way [146]. In this technique, the AgAuNPs formation seems to be independent of the deposition order of the initial metallic layers on the glass substrate. Besides, in the case of the AuAgNPs elliptical formation with metal evaporation, the LSPR band shift is dependent on the shape, i.e., on the degree of sphericity of the AgAuNPs. AgAuNPs were also synthesized by UV laser radiation in the near-surface region of silicate glass [147]. Finally, nanosphere lithography allows the formation of very ordered arrays of silver-gold alloy nanoprisms (AgAuNPrs) on glass support. It has been shown that AgAuNPs are about four times more sensitive in RI than the equivalent supported spherical AgAuNPs with similar sizes and conditions. Moreover, the RIS of the AgAuNPs with $x_{Au} = 0.5$ is closer to pure AgNPs, i.e., very superior to pure AuNPs RIS [150].

AgAu alloys are more stable than gold-silver core@shell nanoparticles for the same size and shape [159]. The composition and the molar ratio between the two metals are important factors to be considered regarding their properties. Indeed, the plasmon peak for spherical AgNPs and AuNPs is around 400 and 520 nm, respectively, while the absorption of AgAuNPs can be tuned continuously from 400 to 520 nm *via* changing the alloy composition [34,160,161]. Qi et al. showed that the alloy NPs became less stable when Ag molar ratio increased conversely to core@shell NPs [159]. Hence, optical properties of alloys mainly depend on the ratio of one metal compared to the other, their size and shape. Indeed, AgAuNPs present only one peak, whose intensity and position in wavelength depend on the molar fraction of Au, x_{Au} [144,161]. In the case of spherical NPs of 18 nm average size, Link et al. demonstrated that the single LSPR peak of the AgAuNPs shifted from 400 nm (pure AgNPs) to 520 nm (pure AuNPs) according to the increasing gold molar fraction [144]. Indeed, it has theoretically predicted and experimentally observed that the LSPR band shift from 400 nm to 520 nm is linear, and proportional to the mole fraction of gold x_{Au}. These results are shown in Figure 17. Moreover, theoretical simulations showed that the intensity of the peak decreases when the gold mole fraction increases [53,145].

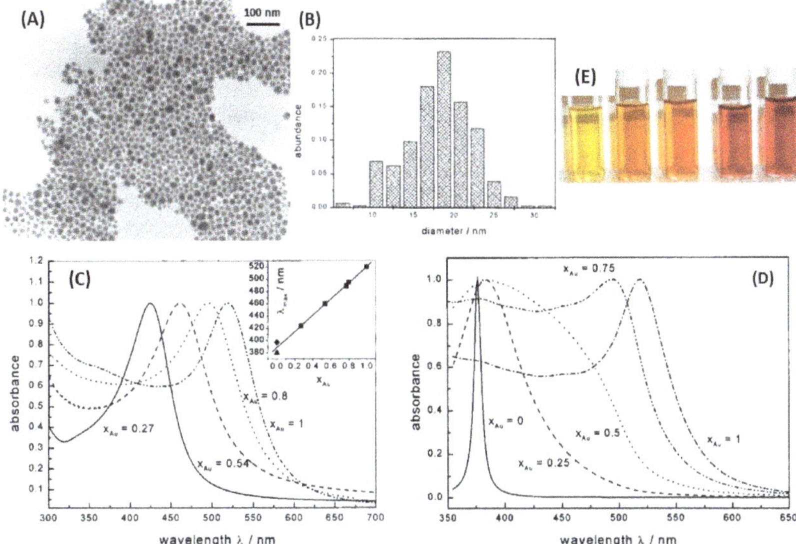

Figure 17. (**A**) TEM image and (**B**) size histogram of spherical AgAuNPs with Au mole fraction x_{Au} = 0.8: the average size is 18 nm. (**C**) Experimental and (**D**) calculated spectra regarding the LSPR shift of 18 nm diameter spherical AgAuNPs with varying Au molar fraction. (**E**) Colloidal suspensions of AuAgNPs with increasing Au concentration. Adapted from [92,144]. Copyright (1999), American Chemical Society. Copyright (2004), Elsevier B.V. All rights reserved.

The application of AgAuNPs to plasmonic biosensing is therefore based on the higher RIS of alloy NPs, compared to pure nanoparticles with equivalent size and shape. Indeed, BSA can be detected with AgAuNPs because a linearity between the LSPR red-shift and the concentration of BSA is observed for concentration between 0.1 and 100 ng/mL, which is better than pure AuNPs [146]. The sensitivity (i.e., the LSPR band shift) and the linearity can even be improved, with use of dopamine coated AgAuNPs.

4.2. Silver and Gold Core@Shell Nanoparticles

Silver and gold core@shell NPs are made of two spatially distinct layers, each containing a different element: a core, made of the first metal (Ag or Au), and a shell made of the second (Au or Ag, respectively). The core@shell notation places core first, thus, Ag@AuNPs refers to silver core coated by a gold shell and vice versa. The shell formation keeps the initial shape of the NPs, imposed by the metallic core, whatever isotropic or anisotropic [162]. The gold shell has essentially a protecting role on Ag core (Ag@AuNPs), to ensure the chemical stability of the previously synthesized AgNPs, and thus is often very thin. In this case, Au electrodeposition on Ag core is the main technique used with the possibility of successive voltammetric cycles to increase the thickness of the Au shell [151,152,157]. Other techniques exist, such as laser ablation of Au in a solution of Ag colloids, to generate the Ag core, on which the Au layer grows (Figure 16A) [53,145]. Here, the growth of the Au shell thickness is followed by UV-Vis spectroscopy, at different ablation times, according to the Mie theory, which makes the link between shell thickness and LSPR peak position. Chemical reduction of $HAuCl_4$ at the AgNPls surface can also lead to very thin Au shells, by adding very slowly the gold solution [153]. The main problem in these techniques is to avoid galvanic replacement of Ag by Au, which would destroy at least partially the Ag core. Indeed, Ag is a more reductive metal than Au. Recently, our group introduced an original pathway to form Ag@AuNPs from hollow gold nanoshell (AuNS) [163]. Porous AuNS were prepared by galvanic replacement starting from AgNPs generating Ag^+ ions in the process (Section 4.3 see below). Increase of pH in the presence of these AuNS triggers the reduction of Ag^+ that preferentially occurs at the inner walls of AuNS. The reaction initially relies on the presence of residual Ag^+ inside the AuNS as well as in the surrounding solution, and it proceeds upon external addition of Ag^+ until a solid Ag core is formed inside the AuNS to form Ag@AuNPs (Figure 18). Then, subsequent reduction of Ag^+ occurs on the external surface of the solidified AuNS (Ag@Au@AgNPs). Controlling the Ag content in AuNS allows tuning the LSPR band position at the desired wavelength for biosensing applications.

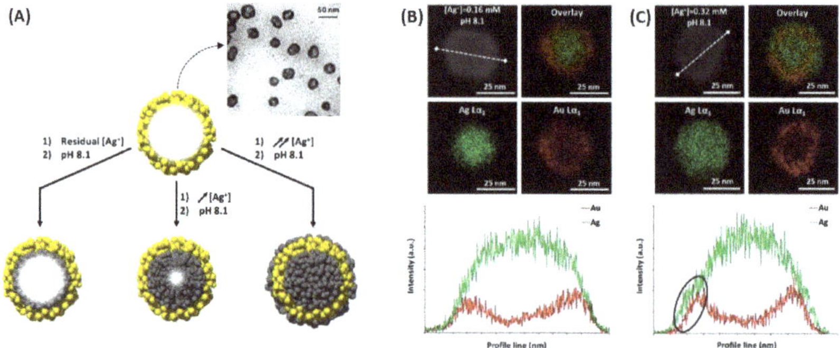

Figure 18. (**A**) Illustration of the reduction and growth process of Ag on the inner and outer surfaces of porous gold nanoshell (AuNS) with increasing amounts of Ag^+ in the surrounding medium. STEM elemental mapping (Ag, Au, and overlay) of AuNS obtained after adding Ag^+: (**B**) $[Ag^+]$ = 0.16 mM and the corresponding elemental profile along the white hatched line and (**C**) $[Ag^+]$ = 0.32 mM and the corresponding elemental profile along the white hatched line. The black ellipse in (**C**) highlights the reduction and growth of Ag at the external surface once the inner volume is completely filled. Adapted from [163]. Copyright (2019), American Chemical Society.

Regarding the synthesis of Au@AgNPs, the chemical reduction of $AgNO_3$ at the AuNPs surface is the main technique used. This requires a reducing agent, which is very often citrate/ascorbate, to form spherical and rod-shaped core@shell structures by Ag chemical deposition on Au core [154–156,161,162]. Besides, a very thin Au layer can be electrodeposited after the Au@AgNPs synthesis, making a peculiar structure, called Au@Ag@AuNPs [157]. Liz-Marzán et al. also realized successive reduction of $AgNO_3$ and $HAuCl_4$ in the presence of AA and CTAB on preformed Au seeds to obtain multilayer bimetallic nanoparticles (Au@Ag, Au@Ag@Au, and Au@Ag@Au@Ag NPs) [164]. According to the Mie theory, for isotropic nanoparticles due to the hybridization between two different plasmonic nanoparticles, the LSPR spectrum should display two peaks, one coming from the core@shell interface between the two metals, and the other one coming from the surface of the shell [53,165]. The position of the former peak mainly depends on the core metal, while the position of the latter one mainly depends on the shell metal, but also on the thickness of the shell layer. As described previously, pure AgNPs have an LSPR extinction peak around 400 nm, and more intense than pure AuNPs. It can be expected that the optimal LSPR properties should occur for very thin Au shells regarding Ag@AuNPs, whereas the Ag shell can be thicker for Au@Ag NPs. Indeed, Zhu et al. demonstrated that the peak of Ag@Au nanowires red-shifted and the intensity decreased when the Au shell thickness increased. In addition, the shell peak is almost inexistent for low Au shell thicknesses. While the peak blue-shifted when the thickness of Ag shell was increased, and the shell peak was more intense for Au@Ag nanowires [166].

Besides, the anisotropic core@shell NPs such as NRs, for which the synthesis was widely described [161,162], the peaks coming from the core@shell interface and the shell surface were enhanced because of the presence of various favored directions, but not all the resulting peaks were always observable. Indeed, four peaks should be observed for Au@AgNRs but only three were actually observed because the two initial peaks were split due to the presence of two favored directions [155].

Moreover, only two remained observable when Ag shell thickness increased. These two peaks corresponded to the longitudinal resonances of the Ag external shell for the shorter wavelength, and Au-Ag interface for the higher wavelength, as in the case of spherical Au@AgNPs [155,167]. The most intense peak observed corresponding to the longitudinal resonance of the Au-Ag interface blue-shifted when the Ag shell thickness increased [156,168]. However, no linearity was observed for Au@AgNRs between the LSPR shift and the Ag shell thickness from 740 nm (Ag shell: 0 nm) to 507 nm (Ag shell: 6 nm), based on the calculated spectra [154]. Figure 19 shows the results in terms of LSPR band position for different Ag shell thickness.

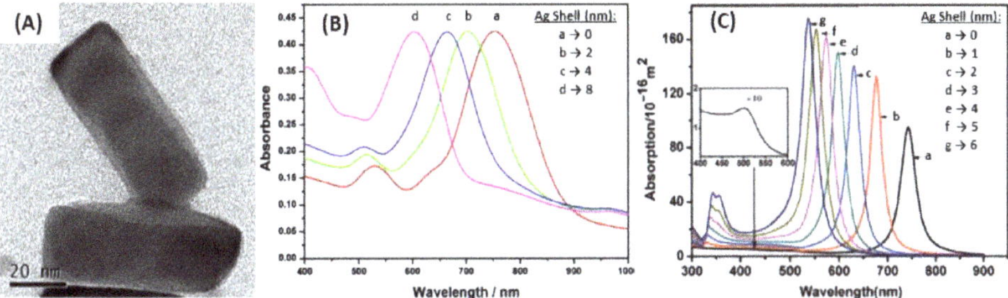

Figure 19. (**A**) TEM image of Au@Ag nanorods with 60×20 nm dimensions for the Au core and 4 nm thickness for the Ag shell and (**B**) variation of the extinction spectra of Au@Ag nanorods with varying Ag shell thickness (0–8 nm) on the 60×20 nm Au core. (**C**) The variations in the calculated LSPR spectra of Au@AgNRs with varying Ag shell thickness (0–6 nm) as well as the zoom on the spectrum allowed seeing the peak corresponding to the Au-Ag interface transversal resonance. Adapted from [154,156]. Copyright (2014), American Scientific Publishers. Copyright (2014), Springer-Verlag Wien.

Several studies pointed out the interesting properties of the core@shell structures based on Ag and Au for LSPR biosensing, because of their high RIS [155]. The LSPR band red-shift is observed when the surrounding RI increases for both Ag@Au and Au@Ag structures [152,156]. Moreover, considering one LSPR peak, the shift in wavelength is proportional to the RI, which is useful for detection of biomolecules in solution [157]. The RIS of core@shell NPs is also dependent on the size and the shape: anisotropic core@shell nanoparticles (NRs, TNPls, etc.) are more sensitive [38] than alloy NPs [169] or anisotropic and isotropic pure NPs [156]. As it has already been described, in the extinction spectrum, there is both an influence of the core/shell interface (depending on the two metals, as in alloys) and shell thickness (depending on the shell metal, as in pure NPs). Additionally, it is possible to deposit a dielectric layer at the surface of Au@Ag, to improve the whole stability as Ag is less stable than Au [170]. In this case, it has been shown that the LSPR sensitivity is not lost, but it can even be raised. Indeed, the LSPR band position increases when the permittivity of the layer is higher than the permittivity of the surrounding medium. The interesting LSPR properties of the core are kept if the protecting layer, which improves stability, is very thin, compared to the core size [151]. Moreover, Dong and his co-workers showed that a certain number of cycles is required for the best efficiency, both for the homogeneity in size and shape and for the RIS in the case of successive Au deposition on Ag core with voltammetric cycles [152,157]. A similar observation was realized for the chemical reduction of Ag on Au, for which a certain quantity of $AgNO_3$ is required for the best sensitivity in RI [156]. Figure 20 shows the linearity between LSPR peak shift and RI, as well as the higher sensitivity of Au@AgNRs compared to Au@AgNPls. A study has also showed Au@Ag core–shell nanorods have better SERS responses, compared AuNRs [154]. Indeed, the SERS intensities increased with the increase of the Ag shell thickness, which demonstrates that the composition and morphology of NPs play key roles on the SERS signals.

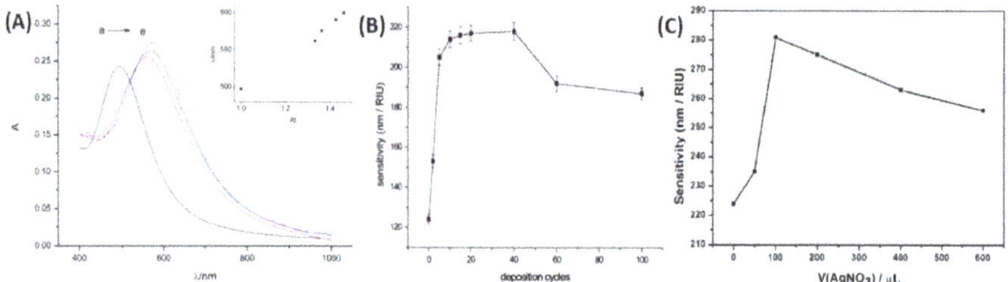

Figure 20. (**A**) Localized surface plasmon resonance (LSPR) band shifts for Ag@Au hemispherical nanoplates (40 nm radius) supported on ITO glass with increasing RI media (a: air, b: water, c: ethanol, d: cyclohexane, e: carbon tetrachloride), and the linear relation between shift and RI (inset). (**B**) Evolution of the RIS with the number of Au shell electrodeposition cycles on the Ag core for the previous Ag@Au nanoplates. (**C**) The evolution of the RIS with the concentration of AgNO₃ for the deposition of the Ag shell on the 20 × 60 nm Au core in the case of Au@AgNRs (TEM image was previously shown in Figure 19A). Adapted from [152,156]. Copyright (2013), American Chemical Society. Copyright (2014), Springer-Verlag Wien.

4.3. Destructive Use of Silver Nanoparticles with Gold

AgNPs can be used as sacrificial template in the destructive way for the synthesis of gold nanobowls [171], gold nanocages (AuNC) [172,173] or nanoshells (AuNS) (Figure 21A) [79,163,174,175]. Indeed, these structures are synthesized by a galvanic replacement reaction, where the metallic salt with higher reduction potential is added to a suspension containing a metal nanoparticle with lower reduction potential, as the following reaction [174,176]:

$$Au^{3+}_{(aq)} + 3\,Ag_{(NPs)} \rightarrow 3\,Ag^+ + Au_{(NC/NS)}$$

Specifically, the standard reduction potential of Au^{3+}/Au redox pair is 0.99 V vs. the standard hydrogen electrode (SHE), whereas the standard reduction potential of Ag^+/Ag is 0.80 V vs. SHE [163]. The difference in reduction potential causes Au to be deposited on the Ag template upon release of Ag^+ into the solution. This method allows designing the shape of the NC or NS as the complementary shape of AgNPs. The nanoshell geometry is ideal for tuning and optimizing the near-field response for SERS on substrates and optical resonance properties of biosensors (Figure 21B) [76,79,174,175]. Tuning the LSPR band of nanoshells into the NIR spectral range leads to a variety of bioapplications.

Regarding the AuNC, the synthesis is done in two steps. The first one is the AgNPs synthesis by electrodeposition on a glass support. Then, the second step consists in the galvanic replacement of Ag by Au, at the AgNPs surface, which can be followed by UV-Vis, using the variation of intensity of the LSPR Au (increasing) and Ag (decreasing) bands, and by cyclic voltammetry. For complete removal of Ag, i.e., to complete gold-silver dealloying, it is necessary to use oxidizing agents such as nitric acid (HNO₃) or hydrogen peroxide (H₂O₂) [173]. Another way to synthesize AuNC is to start from Ag disks deposited on a glass with colloidal lithography, followed by galvanic replacement of Ag by Au to create peculiar AuNC, which are AuAg nanobowls [171].

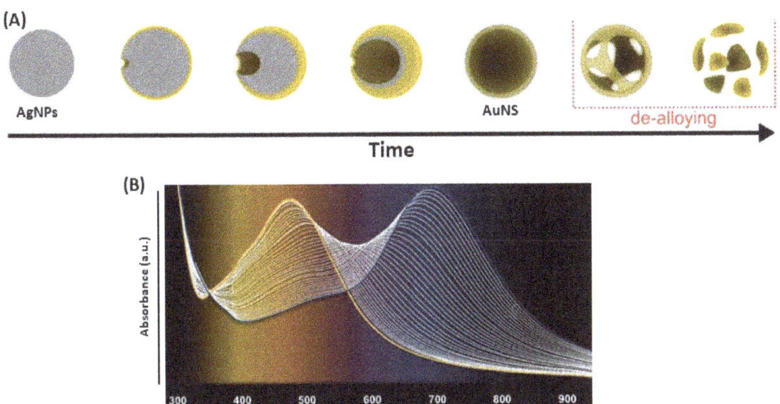

Figure 21. (**A**) The structural evolution of AuAg nanostructures during the galvanic replacement reaction upon addition of $HAuCl_4$ and (**B**) absorption spectra evolution as a function of time of AgNPs titrated with increasing volumes of $HAuCl_4$ to form AuNS: the LSPR band gradually shifts through the whole visible spectrum toward NIR wavelengths. Adapted from [175]. Copyright (2018), American Chemical Society.

5. Selected Applications of Ag and AgAu-Based Plasmonic Nanoparticles in Optical Biosensing

Few studies have reported the use of Ag and AgAu-based plasmonic nanoparticles for biosensor applications. In this last part, selected examples of the use of Ag and Ag-Au nanoparticles in the development of plasmonic biosensors based on biomolecules recognition will be detailed.

5.1. RI-Based LSPR Biosensors

Although AgNPs have been used less extensively than AuNPs in the development of biosensors, very interesting works have been published in LSPR optical biosensing. Indeed, a study showed the use of AgNPs exhibited better results for RI-based LSPR biosensing compared to AuNPs, as discussed previously in the manuscript. The LSPR band shift resulting from the addition of biocytin-coated metallic nanospheres by addition of avidin was approx. 5 times higher for AgNPs than for AuNPs, 1.78 nm/nM vs. 10.18 nm/nM for AuNPs and AgNPs, respectively (Figure 15C) [143]. Another study investigated the development of Ag triangular plasmonic NPs on glass substrate, fabricated by NSL, to lead to sensitive and selective nanoscale affinity biosensors for the streptavidin-biotin couple [57]; the limit of detection (LOD) for these LSPR biosensors was found to be in the low-picomolar to high-femtomolar region (Figure 22A–C). A method to amplify the wavelength shift, previously observed, from LSPR bioassays was optimized using Au nanoparticle-labeled antibiotin antibodies. After binding an antigen to the antibody-conjugated Ag nanotriangles, a secondary antibody attached to AuNPs was added. The resulting plasmonic coupling between the Ag nanotriangles and the Au colloids reduced the LOD by three orders of magnitude for more sensitive detection (Figure 22D–F) [177].

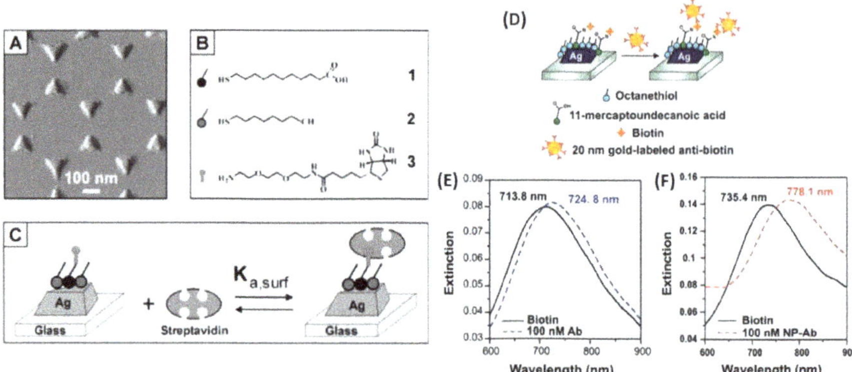

Figure 22. Silver triangular nanoparticles fabricated by NSL on a glass substrate. (**A**) Tapping mode AFM image of the Ag triangular NPs. (**B**) Surface chemistry of the Ag nanobiosensor. A mixed monolayer of (1) 11-MUA and (2) 1-OT is formed on the exposed surfaces of the AgNPs followed by the covalent linking of (3) biotin to the carboxyl groups of (1) 11-MUA. Schematic illustration of (**C**) streptavidin binding to a biotinylated Ag nanobiosensor and (**D**) biotin covalently linked to the Ag nanobiosensor surface while antibiotin-labeled AuNPs are subsequently exposed to the surface. LSPR spectra (**E**) before (solid black) and after (dashed blue) binding of native antibiotin and (**F**) before (solid black) and after (dashed red) binding of antibiotin-labeled NPs. Adapted from [57,177]. Copyright (2002, 2011), American Chemical Society.

To enhance the sensitivity of the LSPR optical sensor, a new and recent approach used by depositing NPs on an optical fiber. The principle of LSPR optical fiber sensors is also based on the plasmon resonance of metal NPs, but coated on optical fiber surfaces, that are more sensitive to changes in the surrounding medium [178,179]. The label-free and real time detection proposed by this technology is a valuable asset compared to classical techniques. However, there are few studies regarding the development of LSPR optical fiber sensors with AgNPs, although nanosilver films have been proven to be much more sensitive to surrounding medium changes than other metal films [180]. Among these studies, Chen et al. proposed a stable and sensitive reflective LSPR optical fiber sensor based on AgNPs to optimize the fabrication process, including two key parameters (the sensing length and the coating time) [179]. The surface of AgNPs deposited on the optical fibers was then functionalized with an antibody and the antigen-antibody binding process was optically monitored by measuring the wavelength shift in real time (Figure 23). This technique gave a RIS of 387 nm/RIU, which is much higher than that reported for colloidal suspension of AgNPs. Another study found a RIS of 67.6 nm/RIU by photodepositing of AgNPs on the optical fiber end [178]. The sensor response is such that the LSPR peak wavelength is linearly shifted to longer wavelength as the RI is increased.

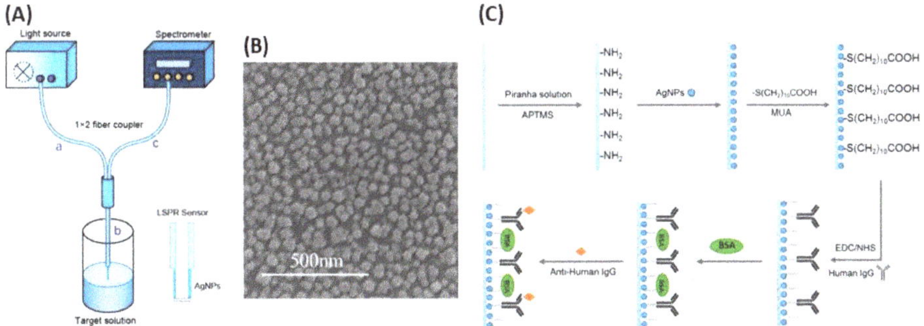

Figure 23. (**A**) Schematic illustration of the experimental set-up used for the LSPR optical fiber sensor. (**B**) SEM image of immobilized AgNPs on optical fiber surface. (**C**) Illustration of the employed strategy for the development of LSPR optical fiber biosensors based on AgNPs. Adapted from [179].

Besides, Patora and Astilean developed LSPR biosensors based on chitosan-coated AgNPs to devise a multi responsive plasmonic sensor [91]. They exploited the anisotropic AgNPs as LSPR chemosensors and *p*-ATP as the target. They showed a gear of plasmon resonance peak, which allows a greater shift toward higher wavelengths. In this same study, chitosan-coated NPs were also used as LSPR sensors for monitoring trace amount of adenine by shifts of LSPR bands proving the binding between the particles and adenine, showing that, the chitosan coated AgNPs make sensitive LSPR sensors and good SERS substrates.

The applications of core@shell NPs combining Ag and Au for RI-based LSPR biosensors are also presented. A study showed that the optical properties of the Au@Ag core@shell NPs were similar to those of pure AgNPs for a given sizes, which was confirmed by means of Mie extinction calculations, while the SERS properties of Au@AgNPs exhibited a higher efficiency than AgNPs under near-infrared excitation [181]. Moreover, the results of three studies focusing on the detection of streptavidin (SA) in solution are briefly discussed below, as a model of optical biosensor, based on the RIS of glass-supported core@shell NPs, and using the receptor-analyte recognition (Figure 24A,B). Biotin, which interacts strongly with the SA target molecule, is fixed on the external shell of the core@shell NPs, which was previously amine-functionalized with 3-aminopropyltrimethoxysilane (APTMS). Two studies used Au@Ag structures, and one Ag@Au structure. Indeed, the SA detection in solution was proven with the use of Au@AgNRs [156], Au@Ag triangular nanoprisms (TNPrs) and Au@Ag@AuTNPrs [157] as well as Ag@Au hemispherical NPls [152]. The results are very comparable: successive shifts in the LSPR peak are observed upon successive additions of APTMS, biotin and SA, in correlation with the induced changes in RI. In addition, linearity is always observed between the LSPR peak shift and SA concentration, as a result of change of local RI. Figure 24C–H shows that the results for SA detection are very similar for Ag@Au and Au@Ag structures, except that the position of the main LSPR peak is red-shifted with Au@Ag core@shell. The Au@AgNRs being more sensitive to RI change, are also more sensitive with respect to SA detection [156] than NPls. Besides, the Au@AgTNPrs have been further coated with a very thin layer of Au. The resulting (Au@Ag@AuTNPrs) keep the initial sensitivity properties of the Au@AgTNPrs, and the linearity between LSPR peak shift and SA concentration [157]. Regarding Ag@Au hemispherical NPls, a complementary study on the biodetection of immunoglobulin G with anti-immunoglobulin G bound Ag@Au NPs, showed similar results to those with SA [152].

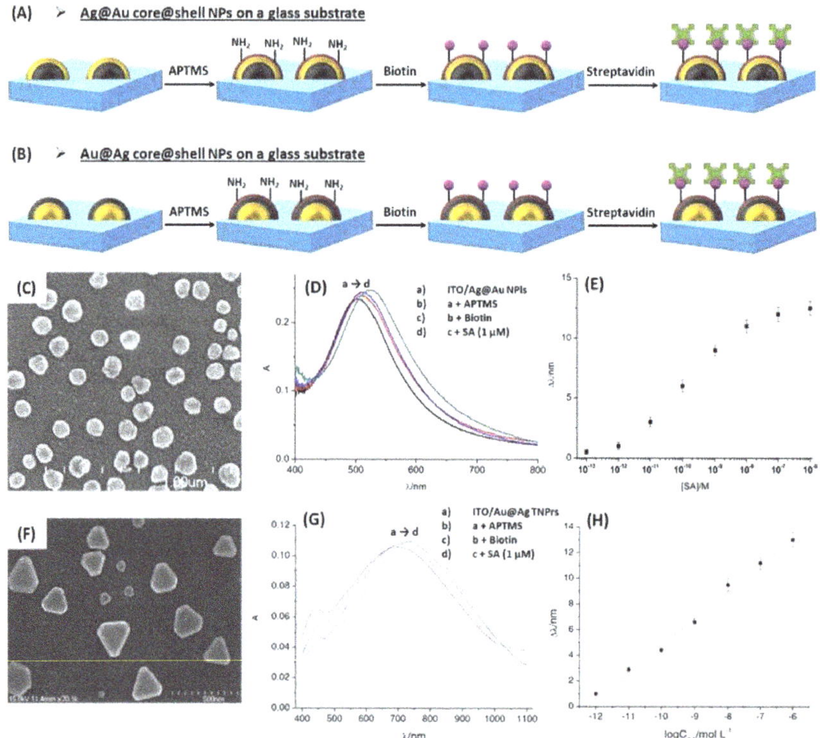

Figure 24. (**A,B**) Schematic illustration showing the preparation of glass-supported core@shell NPs for SA biosensing. (**C**) SEM image of Ag@Au hemispherical nanoplates supported on ITO glass. (**D**) LSPR peak (500 nm) was shifted upon successive treatments with APTMS, biotin and SA. (**E**) Relationship between the LSPR band shift and SA concentration for Ag@Au NPls. (**F**) SEM image of Au@AgTNPls supported on ITO glass. (**G**) LSPR peak (700 nm) shifted upon successive treatment with APTMS, biotin and SA. (**H**) Linear relationship between the LSPR band shift and SA concentration. Adapted from [152,157]. Copyright (2013), American Chemical Society. Copyright (2013), Springer Science Business Media New York.

Finally, the fabrication of substrate-bound AuAg nanobowls arrays synthesized through the galvanic replacement of silver disk arrays is used for size-selective LSPR biosensors. This sensor should prove useful in both size determination and differentiation of large analytes in biological solutions, such as viruses, fungi, and bacterial cells. In these devices, both the LSPR and the SERS signals are enhanced, and the LSPR peak is red-shifted, when the target analyte is small enough to penetrate inside AuAg nanobowls. Otherwise, the previous described effects on LSPR and SERS are not observed [171]. Therefore, this concept was applied towards the detection of a 95 nm H1N1 virus, where the larger diameter nanobowls showed an increased plasmonic response upon addition of the virus.

5.2. Ag and Mixed AgAu Nanoparticle-Based Colorimetric Biosensors

This approach received considerable attention in the analytical field for naked-eye detection due to its simplicity and low cost, it does not require any expensive or complex instrumentation. Due to these inherent optical properties, colloidal suspensions of AgNPs and mixed Ag and Au NPs have high extinction coefficients and different colors in the visible region of the spectrum when they are dispersed in comparison with when they are aggregated.

5.2.1. Ag Nanoparticles Aggregation-Based Colorimetric Assays

In literature, the development of nanoparticle aggregation based-colorimetric assays has been reported. The optical plasmon properties of AgNPs depend strongly on the interparticle distance between pairs of NPs, small or large aggregates of AgNPs as compared to individual and well-spaced NPs. A decrease in the interparticle distance leads to a strong overlap between the plasmon fields of the nearby particles, causing a redshift in the LSPR band with an increase in intensity and an easily observable change in color solution. Indeed, the analytical performance with high sensitivities because of the strong LSPR and excellent selectivity driven by the interaction between analyte-NPs and its surroundings involving mainly electrostatic and hydrogen bond interactions as well as donor–acceptor chemical reactions. Therefore, a well-designed chemical interaction could lead to a change of color for naked-eye detection of the target analyte [182]. AgNPs-based colorimetric assays have been investigated for melamine detection [183,184]. Han et al. used *p*-nitroaniline (*p*-NA)-modified AgNPs, as a sensitive, selective and simple colorimetric assay, resulting in a color change from yellow to blue in the presence of melamine [183]. This optical method was highly reproducible and concentrations as low as 0.1 ppm of melamine in infant formula can be visualized by the naked-eye. The same strategy was proposed by Ma et al. with dopamine-stabilized AgNPs to detect visually the melamine [184]. Indeed, the color change of the suspension turned from yellow to brown (Figure 25). The results showed the concentrations of detectable melamine were in the range of 10 ppb to 1.26 ppm. AgNPs functionalization and the analysis of melamine can occur in one-step since *p*-NA or dopamine acts as a reducer and stabilizer of AgNPs and as a linker of melamine molecules. Other biomolecules has also been detected by a colorimetric sensor highly selective such as tryptophan [185] or histidine [186] thanks to organic coating on AgNPs: 4.4-bipyridine (4-DPD)- and *p*-sulfonatocalixarene (*p*-SC4)-modified AgNPs, respectively. In the first case, tryptophan interacts with the pyridine ring of 4-DPD *via* π-π interactions, and meanwhile carboxyl acid of tryptophan can also form hydrogen bonds with pyridine, which results in 4-DPD-functionalized AgNPs aggregation and the color change from yellow to red. The LOD was 20 µM for the tryptophan colorimetric detection [185]. In the second case, the aggregation process is due to *p*-SC4, which possesses an electron-rich cyclic cavity, being able to fit imidazole and the side chain of histidine *via* host–guest electrostatic and cation-π interactions. The color change turned from yellow to red and the LOD of 5 µM was obtained [186]. Besides, a derivative of calixarene has also been employed for colorimetric detection of pesticides in water [187].

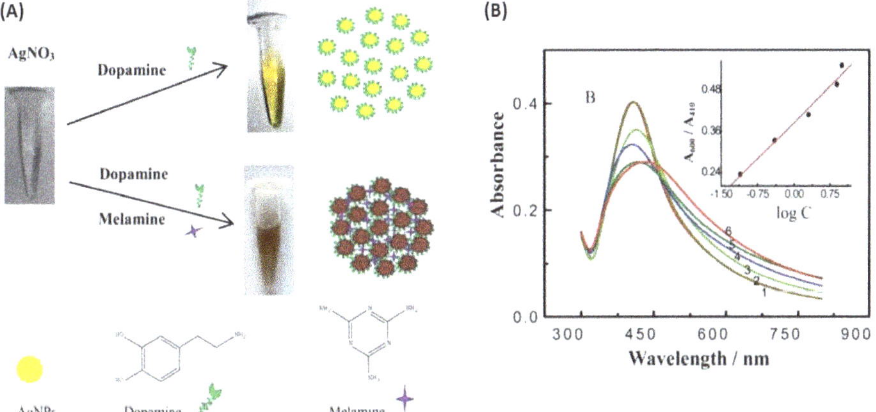

Figure 25. (**A**) Principle of nanoparticle aggregation based-colorimetric assay for the melamine detection with dopamine-modified AgNPs. (**B**) UV-Vis spectra of dopamine-stabilized AgNPs suspensions with different melamine concentrations: (1) 0 mM, (2) 0.08 mM, (3) 0.4 mM, (4) 2 mM, (5) 8 mM and (6) 10 mM. Adapted from [184]. Copyright (2011), Royal Society of Chemistry.

5.2.2. Mixed AgAu Nanoparticles-Based Colorimetric Assays

A colorimetric biosensor with naked-eye detection was designed in which target DNA was indirectly detected through reduction of Ag^+ to AgNPs at the surface of gold nanostars coated by capture probe (Figure 26) [188]. In the presence of target DNA, biotin-labeled complementary oligonucleotide H1, oligonucleotide H2 and avidin-alkaline phosphatase conjugate, ascorbic acid 2-phosphate is converted into AA, which acts as reducer. The resulting observation is a blue-shift of the LSPR peak spectrum, due to the formation of Ag shell on gold nanostars. This technique showed a detection range from 10 fM to 50 pM DNA with a detection limit of 2.6 fM.

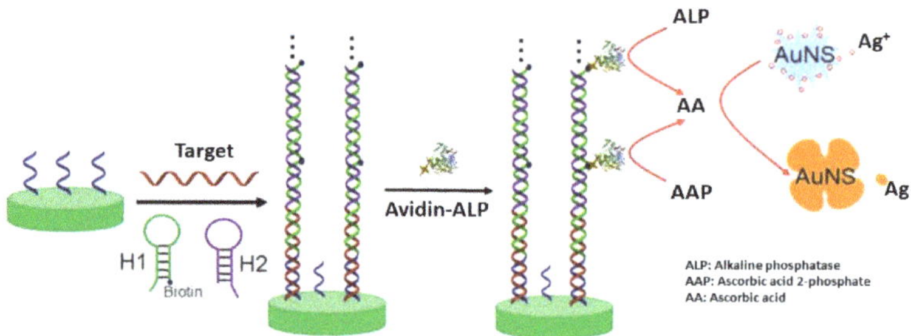

Figure 26. DNA detection by naked-eye readout with the silver reduction on gold nanostars [188]. Copyright (2015), Elsevier B.V. All rights reserved.

A more recent technique based on non-aggregated Au@Ag core@shell NPs was developed by Mao et al. to detect drugs, such as cocaine, using the coloration of the solution containing NPs [189]. For this purpose, they have used Au@Ag nanoparticles coated with a DNA aptamer specific of cocaine and magnetic beads coated with a DNA sequence, partially complementary to the aptamer sequence allowing cross-linking of Au@Ag nanoparticles and magnetic beads. In the presence of a magnetic field, the nanoparticles leave the suspension with the magnetic beads, and the solution is slightly colored. When cocaine is added to the solution, it interacts with the aptamer, destroying the link between nanoparticles and magnetic beads, allowing the coloration of the solution by the nanoparticles (Figure 27). This study shows the new major improvements in the silver plasmonic biosensing.

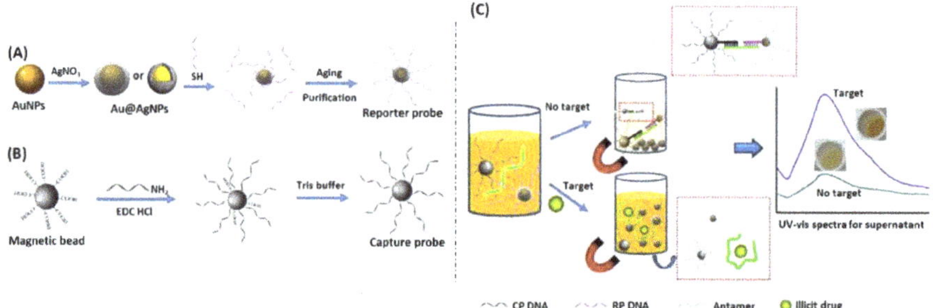

Figure 27. Schematic illustration of the preparation of the (**A**) reporter probe and (**B**) capture probe as well as the principle of the colorimetric detection of illicit drug based on non-aggregation Au@Ag core@shell NPs. Reproduced with permission from [189]. Copyright (2017), Published by Elsevier B.V.

5.3. Metal-Enhanced Fluorescence (MEF)-Based Biosensors

Although fluorescence is a highly sensitive technique, where single molecules can readily be recognized, the detection of a fluorophore is usually limited by its quantum yield, auto-fluorescence of the samples and/or the photo-stability of the fluorophores. However, the use of metallic nanostructures such as silver allows modifying favorably the spectral properties of fluorophores and reducing some of these fluorophore disadvantages for metal-enhanced fluorescence (MEF) [190,191]. An interesting study has reported the use of a relatively facile deposition of AgNPs onto glass slides (i.e., silver island films, SIFs). This sensor was used for the development of an enhanced detection limit sandwich-format immunoassay for the cardiac marker myoglobin (Figure 28A) [191]. Indeed, the SIFs and glass surfaces were coated with anti-myoglobin antibodies, and then incubated with fluorophore-labeled anti-myoglobin antibodies. This approach of metal-enhanced planar immunoassay showed a 10–15-fold increase in fluorescence emission observed on the SIFs compared to that naked substrate (Figure 28B) and the results demonstrated the myoglobin concentrations were detected in the 10–1000 ng/mL range (Figure 28C).

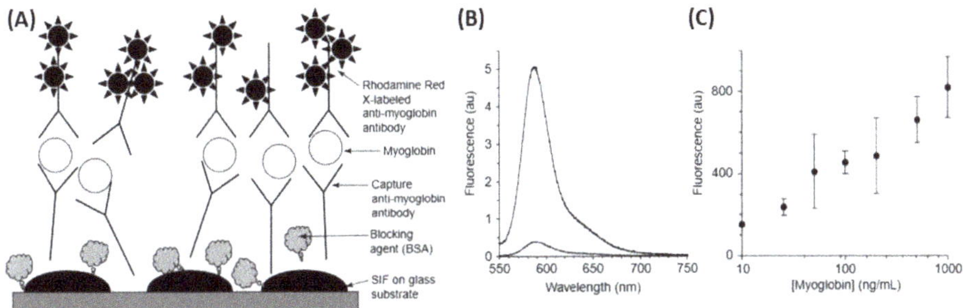

Figure 28. (**A**) Illustration of a metal-enhanced sandwich immunoassay on silver island films (SIFs). Fluorescence emission of the Rhodamine Red-X-labeled anti-myoglobin antibody attached to the surface-immobilized myoglobin (**B**) for a given myoglobin concentration (100 ng/mL) on SIFs and on glass, and (**C**) at different myoglobin concentrations on SIFs. Adapted from [191]. Copyright (2005), Elsevier Ltd. All rights reserved.

AgNPs functionalized with various thiolates have been also explored for MEF-based biosensors [191]. Indeed, DNA hybridization assays using metal-enhanced fluorescence (MEF) were investigated with thiolated oligonucleotides, which were bound to AgNPs on a glass substrate. This approach suggested the use of AgNPs improved the sensitivity of DNA detection with an increase in the number of detected photons per fluorophore molecule by a factor of 10-fold or more [192]. In addition, another study has also used thiol-organic monolayer-protected AgNPs, which were displaced by oligonucleotides through ligand exchanges, and a fluorophore-labeled complementary oligonucleotide were employed for DNA hybridization. The results showed a possible approach to DNA detection with a surface-enhanced emission after hybridization in the presence of AgNPs based on the aggregation of AgNPs bound by fluorophore-labeled oligonucleotides [193]. Finally, Ag@SiO$_2$ NPs have been also exploited as transducers of DNA hybridization [141], or to achieve MEF-based biosensor [194,195]. In the latter case, 3- to 5-fold enhanced fluorescence signals can be obtained from SiO$_2$-coated AgNPs colloids labeled with cyanine and by their aggregation in suspension. This inert coating reduces the close proximity quenching by noble metals, as well as provides for a wide variety of chemistries for biomolecule attachment. AgNPs@SiO$_2$ can thus become a solution-based enhanced-fluorescence sensing platform [190,191,194].

5.4. Optical Biosensors Based on the Oxidation of Ag

The oxidation of silver from AgNPs and mixed AgAu NPs, even in destructive way, can be exploited for optical biosensors. Indeed, a work has used the highly reactive properties of H$_2$O$_2$ to modify the nanoparticle shape for improved detection by colorimetric visualization. Xia et al. used the enzyme glucose oxidase mixed in solution with Ag nanoprisms to catalyze the reaction between glucose and oxygen to form H$_2$O$_2$ and gluconic acid. As the reactive H$_2$O$_2$ etched the tips of the nanoprisms, drastic shape and color changes were observed in the LSPR spectrum, resulting in a detection range from 0.2 μM to 100 μM in diluted serum (Figure 29) [196].

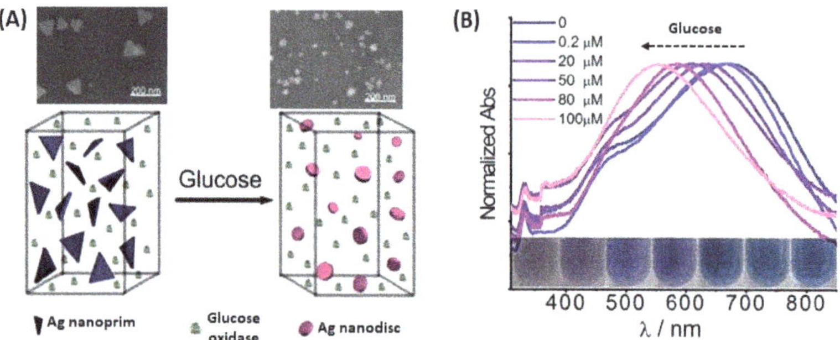

Figure 29. (**A**) Illustration of the strategy to modify the AgNPs shape from nanoprisms to nanodiscs through Ag oxidation for colorimetric sensing of glucose. SEM images of the Ag nanoprisms before and after incubation with glucose oxidase and glucose (100 μM) for 60 min are also showed. (**B**) Absorption spectra of the Ag nanoprisms after glucose incubation in various concentrations for 40 min with photographs of the corresponding suspensions. Adapted from [196]. Copyright (2013), American Chemical Society.

Furthermore, the destructive use of AgNPs is described by Ag oxidation for the synthesis of gold nanocages (AuNC) as well as hollow gold nanoshells (AuNS). First, the dissolution of the Ag part of the NPs is followed by the changes in the initial UV-Vis spectrum, either the shift of the peaks, or the changes in absorbance allow quantifying the presence of the biomolecule [197]. Several studies also demonstrated that glucose can be detected by this technique, using gold-silver nanoshells (AuAgNS) or Au@AgNPs [198,199]. In both cases, Ag is oxidized to Ag^+ because of H_2O_2 produced (Figure 30A) from glucose oxidation to gluconic acid in the presence of dioxygen (O_2) and glucose oxidase. This strategy is based on the quantification of the amount of metallic Ag oxidized, then the quantity of H_2O_2 is determined as well as the quantity of glucose. In 2012, a study on Ag oxidation in presence of glucose with dioxygen and glucose oxidase in AuAgNS showed that the LSPR peak is red-shifted with increasing glucose concentration (Figure 30B) [198]. There is linearity between the wavelength shift and the glucose concentration at very low concentrations (down to 20 μM). In another study, the same oxidation process of Ag in presence of glucose with O_2 was observed in Au@Ag core@shell structures. The LSPR band intensity, at fixed wavelength was correlated to glucose concentration. In this case, a linearity was observed between absorbance and glucose concentration for higher glucose concentrations (down to 400 μM), and a similar observation has been made for cholesterol detection, using the same method of Ag oxidation [199].

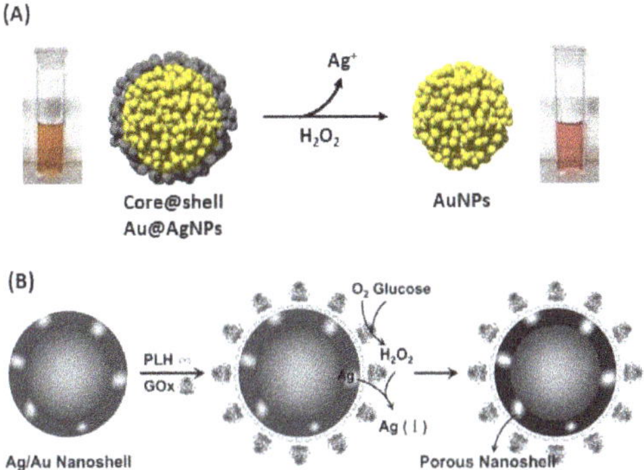

Figure 30. (**A**) Illustration showing the oxidation of Ag shell on Au@Ag core@shell NPs to Ag^+ in the presence of H_2O_2 produced by an enzymatic reaction. (**B**) Schematic illustration showing the glucose sensing mechanism with Ag/Au nanoshells. Adapted from [198]. Copyright (2012) WILEY-VCH Verlag GmbH & Co. KGaA, Weinheim.

6. Conclusions

The interest of silver nanoparticles as highly sensitive materials for plasmonic biosensors design is well-established. Indeed, although AgNPs are less chemically stable and less biocompatible compared to AuNPs, they provide more sensitive plasmonic biosensors owing to their LSPR features. The AgNPs synthesis is now well mastered and well described allowing the fabrication of differently shaped particles from the simplest to special uncommon shapes thanks to the large range of synthesis techniques now available and described in this manuscript for a conceptual opportunity in biosensing. This is a real advantage to explore many more properties. The coating, either organic or inorganic, overcomes the issues of stability and toxicity raised above and allows for the use of the resulting core@shell nanoparticles in plasmonic biosensing. Finally, the use of gold with silver nanoparticles, in alloy and core@shell structures, also provides a protective shell but in addition, enhances the plasmonic response of the resulting colloids.

Most of these findings are recent, and this may explain the few biosensing applications based on AgNPs compared to AuNPs to date. We expect growing interest to the application of these nano-objects in biosensing field, thanks to their higher RIS that allows for a better sensitivity when the strategies are based on the shift of the LSPR band. They are also very promising in naked-eye detection strategies, where multiple scenarios can be envisioned including aggregation, visible lambda shift, or even destruction of a silver shell on a gold core. In this review, the majority of selected applications of Ag and mixed AgAu nanoparticles-based plasmonic biosensors represents only trivial biosensing schemes to emphasize the merit of Ag-related NPs and provide the future prospects silver-based plasmonic nanoparticles in biosensing. In such an application, the expectations for an on-site biosensor, i.e., sensitive, reliable, fast, and user-friendly, are completely fulfilled.

Author Contributions: Conceptualization, S.B. initiated and directed the project—V.A., G.B.-A., and M.L. equally contributed to the original draft preparation, A.L. and S.B. reviewed and edited the original draft. A.L. wrote the manuscript starting from the original draft. S.B. and M.S. supervised and corrected the review. All authors contributed to the final manuscript.

Funding: This research was funded ANR-FWF grant "NanoBioSensor", ANR-15-CE29-0026-02 and by Idex Sorbonne Universities, Form@Innov.

Acknowledgments: The authors thank Sorbonne University and the Master of Chemistry for supporting this initiative.

Conflicts of Interest: The authors declare no conflict of interest.

Abbreviations

AA	Ascorbic acid
Ag	Silver
$AgNO_3$	Silver nitrate
AgNPs	Silver nanoparticles
Ag@AuNPs	Silver gold core@shell nanoparticles
AgAuNPs	Silver gold alloy nanoparticles
APTMS	3-Aminopropyltrimethoxysilane
Au	Gold
AuNPs	Gold nanoparticles
BSA	Bovine serum albumin
CTAB	Cetyltrimethylammonium bromide
Cu	Copper
FDA	Food and Drug Administration
$HAuCl_4$	Tetrachloroauric (III) acid
ITO	Indium tin oxide
LOD	Limit of detection
LSPR	Localized surface plasmon resonance
MEF	Metal-enhanced fluorescence
MOF	Metal-organic framework
MPTES	(3-Mercaptopropyl) triethoxysilane
$NaBH_4$	Sodium borohydride
NaHS	Sodium hydrosulfide
Na_2S	Sodium disulfide
NC	Nanocages
NIR	Near-infrared
NOM	Natural organic matter
NPls	Nanoplates
NPrs	Nanoprisms
NPs	Nanoparticles
NRs	Nanorods
NS	Nanoshells
NSL	Nanosphere lithography
ORG	Oxidation reduction growth
PEG	Poly(ethylene) glycol
PVA	Poly(vinyl acetate)
PVP	Polyvinyl pyrrolidone
RI	Refractive index
RIS	Refractive index sensitivity
RIU	Refractive index unit
SA	Streptavidin
SDS	Sodium dodecyl sulfate
SEM	Scanning electron microscopy
SERS	Surface-enhanced Raman spectroscopy
SIF	Silver island film
SHE	Standard hydrogen electrode
SPP	Surface plasmon polariton
SPR	Surface plasmon resonance
TEM	Transmission electron microcopy
TEOS	Tetraethyl orthosilicate
TNPls	Triangular nanoplates
TNPrs	Triangular nanoprisms
TOAB	Tetraoctylammonium bromide
VdW	Van der Waals

References

1. Alexander, J.W. History of the medical use of silver. *Surg. Infect.* **2009**, *10*, 289–292. [CrossRef]
2. Khaydarov, R.R.; Khaydarov, R.A.; Estrin, Y.; Evgrafova, S.; Scheper, T.; Endres, C.; Cho, S.Y. Silver nanoparticles. In *Nanomaterials: Risks and Benefits*; Springer: Dordrecht, The Netherlands, 2009; pp. 287–297.
3. Castellano, J.J.; Shafii, S.M.; Ko, F.; Donate, G.; Wright, T.E.; Mannari, R.J.; Payne, W.G.; Smith, D.J.; Robson, M.C. Comparative evaluation of silver-containing antimicrobial dressings and drugs. *Int. Wound J.* **2007**, *4*, 114–122. [CrossRef]
4. Marambio-Jones, C.; Hoek, E.M.V. A review of the antibacterial effects of silver nanomaterials and potential implications for human health and the environment. *J. Nanopart. Res.* **2010**, *12*, 1531–1551. [CrossRef]
5. Quang Huy, T.; van Quy, N.; Anh-Tuan, L. Silver nanoparticles: Synthesis, properties, toxicology, applications and perspectives. *Adv. Nat. Sci. Nanosci. Nanotechnol.* **2013**, *4*, 033001. [CrossRef]
6. Prabhu, S.; Poulose, E.K. Silver nanoparticles: Mechanism of antimicrobial action, synthesis, medical applications, and toxicity effects. *Int. Nano Lett.* **2012**, *2*, 32. [CrossRef]
7. Rai, M.; Yadav, A.; Gade, A. Silver nanoparticles as a new generation of antimicrobials. *Biotechnol. Adv.* **2009**, *27*, 76–83. [CrossRef]
8. Nowack, B.; Krug, H.F.; Height, M. 120 years of nanosilver history: Implications for policy makers. *Environ. Sci. Technol.* **2011**, *45*, 1177–1183. [CrossRef]
9. Stockman, M.I. Nanoplasmonics: The physics behind the applications. *Phys. Today* **2011**, *64*, 39–44. [CrossRef]
10. Catherine, L.; Olivier, P. *Gold Nanoparticles for Physics, Chemistry and Biology*; World Scientific: London, UK, 2017. [CrossRef]
11. Purohit, K.; Khitoliya, P.; Purohit, R. Recent advances in nanotechnology. *Int. J. Sci. Eng. Res.* **2012**, *3*.
12. Yu, H.-D.; Regulacio, M.D.; Ye, E.; Han, M.-Y. Chemical routes to top-down nanofabrication. *Chem. Soc. Rev.* **2013**, *42*, 6006–6018. [CrossRef]
13. Chan, H.-K.; Kwok, P.C.L. Production methods for nanodrug particles using the bottom-up approach. *Adv. Drug Deliv. Rev.* **2011**, *63*, 406–416. [CrossRef]
14. Ariga, K.; Yamauchi, Y.; Rydzek, G.; Ji, Q.; Yonamine, Y.; Wu, K.C.W.; Hill, J.P. Layer-by-layer nanoarchitectonics: Invention, innovation, and evolution. *Chem. Lett.* **2013**, *43*, 36–68. [CrossRef]
15. Biswas, A.; Bayer, I.S.; Biris, A.S.; Wang, T.; Dervishi, E.; Faupel, F. Advances in top–down and bottom–up surface nanofabrication: Techniques, applications & future prospects. *Adv. Colloid Interface Sci.* **2012**, *170*, 2–27. [CrossRef]
16. Sanchez, F.; Sobolev, K. Nanotechnology in concrete—A review. *Constr. Build. Mater.* **2010**, *24*, 2060–2071. [CrossRef]
17. Cao, G. *Nanostructures & Nanomaterials: Synthesis, Properties & Applications*; Imperial College Press: London, UK, 2004. [CrossRef]
18. Zhang, X.; Guo, Q.; Cui, D. Recent Advances in nanotechnology applied to biosensors. *Sensors* **2009**, *9*, 1033. [CrossRef]
19. Verma, J.; Lal, S.; van Noorden, C.J. Inorganic nanoparticles for the theranostics of cancer. *Eur. J. Nanomed.* **2015**, *7*, 271–287. [CrossRef]
20. Daniel, M.-C.; Astruc, D. Gold nanoparticles: Assembly, supramolecular chemistry, quantum-size-related properties, and applications toward biology, catalysis, and nanotechnology. *Chem. Rev.* **2004**, *104*, 293–346. [CrossRef]
21. Huang, H.-C.; Barua, S.; Sharma, G.; Dey, S.K.; Rege, K. Inorganic nanoparticles for cancer imaging and therapy. *J. Control. Release* **2011**, *155*, 344–357. [CrossRef]
22. Ghosh Chaudhuri, R.; Paria, S. Core/shell nanoparticles: Classes, properties, synthesis mechanisms, characterization, and applications. *Chem. Rev.* **2012**, *112*, 2373–2433. [CrossRef]
23. Doane, T.L.; Burda, C. The unique role of nanoparticles in nanomedicine: Imaging, drug delivery and therapy. *Chem. Soc. Rev.* **2012**, *41*, 2885–2911. [CrossRef]
24. Dykman, L.; Khlebtsov, N. Gold nanoparticles in biomedical applications: Recent advances and perspectives. *Chem. Soc. Rev.* **2012**, *41*. [CrossRef]
25. Chen, A.; Chatterjee, S. Nanomaterials based electrochemical sensors for biomedical applications. *Chem. Soc. Rev.* **2013**, *42*, 5425–5438. [CrossRef]

26. Anker, J.N.; Hall, W.P.; Lyandres, O.; Shah, N.C.; Zhao, J.; van Duyne, R.P. Biosensing with plasmonic nanosensors. *Nat. Mater.* **2008**, 7, 442. [CrossRef]
27. Mayer, K.M.; Hafner, J.H. Localized surface plasmon resonance sensors. *Chem. Rev.* **2011**, 111, 3828–3857. [CrossRef]
28. Boisselier, E.; Astruc, D. Gold nanoparticles in nanomedicine: Preparations, imaging, diagnostics, therapies and toxicity. *Chem. Soc. Rev.* **2009**, 38, 1759–1782. [CrossRef]
29. Dreaden, E.C.; Alkilany, A.M.; Huang, X.; Murphy, C.J.; El-Sayed, M.A. The golden age: Gold nanoparticles for biomedicine. *Chem. Soc. Rev.* **2012**, 41, 2740–2779. [CrossRef]
30. Rycenga, M.; Cobley, C.M.; Zeng, J.; Li, W.; Moran, C.H.; Zhang, Q.; Qin, D.; Xia, Y. Controlling the synthesis and assembly of silver nanostructures for plasmonic applications. *Chem. Rev.* **2011**, 111, 3669–3712. [CrossRef]
31. Unser, S.; Bruzas, I.; He, J.; Sagle, L. Localized surface plasmon resonance biosensing: Current challenges and approaches. *Sensors* **2015**, 15, 15684. [CrossRef]
32. Hui, S. Plasmonic nanoparticles: Towards the fabrication of biosensors. *IOP Conf. Ser. Mater. Sci. Eng.* **2015**, 87, 012009. [CrossRef]
33. Chen, P.; Tran, N.T.; Wen, X.; Xiong, Q.; Liedberg, B. Inflection Point of the localized surface plasmon resonance peak: A General method to improve the sensitivity. *ACS Sens.* **2017**, 2, 235–242. [CrossRef]
34. Chen, P.; Liu, X.; Goyal, G.; Tran, N.T.; Shing Ho, J.C.; Wang, Y.; Aili, D.; Liedberg, B. Nanoplasmonic Sensing from the human vision perspective. *Anal. Chem.* **2018**, 90, 4916–4924. [CrossRef]
35. Hossain, M.K.; Kitahama, Y.; Huang, G.G.; Han, X.; Ozaki, Y. Surface-enhanced Raman scattering: Realization of localized surface plasmon resonance using unique substrates and methods. *Anal. Bioanal. Chem.* **2009**, 394, 1747–1760. [CrossRef]
36. Tabor, C.; Murali, R.; Mahmoud, M.; El-Sayed, M.A. On the use of plasmonic nanoparticle pairs as a plasmon ruler: The Dependence of the near-field dipole plasmon coupling on nanoparticle size and shape. *J. Phys. Chem. A* **2009**, 113, 1946–1953. [CrossRef]
37. Ray, P.C. Size and shape dependent second order nonlinear optical properties of nanomaterials and their application in biological and chemical sensing. *Chem. Rev.* **2010**, 110, 5332–5365. [CrossRef]
38. Steinbrück, A.; Stranik, O.; Csaki, A.; Fritzsche, W. Sensoric potential of gold-silver core-shell nanoparticles. *Anal. Bioanal. Chem.* **2011**, 401, 1241. [CrossRef]
39. Huang, H.; Huang, S.; Yuan, S.; Qu, C.; Chen, Y.; Xu, Z.; Liao, B.; Zeng, Y.; Chu, P.K. High-sensitivity biosensors fabricated by tailoring the localized surface plasmon resonance property of core-shell gold nanorods. *Anal. Chim. Acta* **2011**, 683, 242–247. [CrossRef]
40. Lu, L.; Burkey, G.; Halaciuga, I.; Goia, D.V. Core-shell gold/silver nanoparticles: Synthesis and optical properties. *J. Colloid Int. Sci.* **2013**, 392, 90–95. [CrossRef]
41. Willets, K.A.; van Duyne, R.P. Localized surface plasmon resonance spectroscopy and sensing. *Ann. Rev. Phys. Chem.* **2007**, 58, 267–297. [CrossRef]
42. Gopinath, S.C.B. Biosensing applications of surface plasmon resonance-based Biacore technology. *Sens. Actuators B Chem.* **2010**, 150, 722–733. [CrossRef]
43. Mortazavi, D.; Kouzani, A.Z.; Kaynak, A.; Duan, W. Nano-plasmonic biosensors: A review. In Proceedings of the 2011 IEEE/ICME International Conference on Complex Medical Engineering, Harbin, China, 22–25 May 2011; pp. 31–36.
44. Sharma, B.; Frontiera, R.R.; Henry, A.-I.; Ringe, E.; van Duyne, R.P. SERS: Materials, applications, and the future. *Mater. Today* **2012**, 15, 16–25. [CrossRef]
45. Eustis, S.; El-Sayed, M.A. Why gold nanoparticles are more precious than pretty gold: Noble metal surface plasmon resonance and its enhancement of the radiative and nonradiative properties of nanocrystals of different shapes. *Chem. Soc. Rev.* **2006**, 35, 209–217. [CrossRef]
46. Lee, K.-S.; El-Sayed, M.A. Gold and Silver nanoparticles in sensing and imaging: Sensitivity of plasmon response to size, shape, and metal composition. *J. Phys. Chem. B* **2006**, 110, 19220–19225. [CrossRef]
47. Huang, X.; El-Sayed, M.A. Gold nanoparticles: Optical properties and implementations in cancer diagnosis and photothermal therapy. *J. Adv. Res.* **2010**, 1, 13–28. [CrossRef]
48. Jain, P.K.; Lee, K.S.; El-Sayed, I.H.; El-Sayed, M.A. Calculated absorption and scattering properties of gold nanoparticles of different size, shape, and composition: Applications in biological imaging and biomedicine. *J. Phys. Chem. B* **2006**, 110, 7238–7248. [CrossRef]

49. Elghanian, R.; Storhoff, J.J.; Mucic, R.C.; Letsinger, R.L.; Mirkin, C.A. Selective colorimetric detection of polynucleotides based on the distance-dependent optical properties of gold nanoparticles. *Science* **1997**, *277*, 1078. [CrossRef]
50. Rosi, N.L.; Mirkin, C.A. Nanostructures in biodiagnostics. *Chem. Rev.* **2005**, *105*, 1547–1562. [CrossRef]
51. Murphy, C.J.; Gole, A.M.; Hunyadi, S.E.; Stone, J.W.; Sisco, P.N.; Alkilany, A.; Kinard, B.E.; Hankins, P. Chemical sensing and imaging with metallic nanorods. *Chem. Commun.* **2008**, *8*, 544–557. [CrossRef]
52. Sepúlveda, B.; Angelomé, P.C.; Lechuga, L.M.; Liz-Marzán, L.M. LSPR-based nanobiosensors. *Nano Today* **2009**, *4*, 244–251. [CrossRef]
53. Navas, M.P.; Soni, R.K. Laser-generated bimetallic Ag-Au and Ag-Cu core-shell nanoparticles for refractive index sensing. *Plasmonics* **2015**, *10*, 681–690. [CrossRef]
54. Khodashenas, B.; Ghorbani, H.R. Synthesis of silver nanoparticles with different shapes. *Arab. J. Chem.* **2015**, 1–16. [CrossRef]
55. Mahmudin, L.; Suharyadi, E.; Utomo, A.B.S.; Abraha, K. Optical properties of silver nanoparticles for surface plasmon resonance (SPR)-based biosensor applications. *J. Mod. Phys.* **2015**, *6*, 1071. [CrossRef]
56. Cobley, C.M.; Skrabalak, S.E.; Campbell, D.J.; Xia, Y. Shape-controlled synthesis of silver nanoparticles for plasmonic and sensing applications. *Plasmonics* **2009**, *4*, 171–179. [CrossRef]
57. Haes, A.J.; van Duyne, R.P. A nanoscale optical biosensor: Sensitivity and selectivity of an approach based on the localized surface plasmon resonance spectroscopy of triangular silver nanoparticles. *J. Am. Chem. Soc.* **2002**, *124*, 10596–10604. [CrossRef] [PubMed]
58. Stavytska-Barba, M.; Salvador, M.; Kulkarni, A.; Ginger, D.S.; Kelley, A.M. Plasmonic enhancement of Raman scattering from the organic solar cell material P3HT/PCBM by triangular silver nanoprisms. *J. Phys. Chem. C* **2011**, *115*, 20788–20794. [CrossRef]
59. Haes, A.J.; Chang, L.; Klein, W.L.; van Duyne, R.P. Detection of a biomarker for Alzheimer's disease from synthetic and clinical samples using a nanoscale optical biosensor. *J. Am. Chem. Soc.* **2005**, *127*, 2264–2271. [CrossRef] [PubMed]
60. Elfassy, E.; Mastai, Y.; Salomon, A. Cysteine sensing by plasmons of silver nanocubes. *J. Solid State Chem.* **2016**, *241*, 110–114. [CrossRef]
61. Merk, V.; Nerz, A.; Fredrich, S.; Gernert, U.; Selve, S.; Kneipp, J. Optical properties of silver nanocube surfaces obtained by silane immobilization. *Nanospectroscopy* **2014**, *1*. [CrossRef]
62. Cheng, Y.; Wang, R.; Zhai, H.; Sun, J. Stretchable electronic skin based on silver nanowire composite fiber electrodes for sensing pressure, proximity, and multidirectional strain. *Nanoscale* **2017**, *9*, 3834–3842. [CrossRef]
63. Yun, H.J.; Kim, S.J.; Hwang, J.H.; Shim, Y.S.; Jung, S.-G.; Park, Y.W.; Ju, B.-K. Silver nanowire-IZO-conducting polymer hybrids for flexible and transparent conductive electrodes for organic light-emitting diodes. *Sci. Rep.* **2016**, *6*, 34150. [CrossRef]
64. Hu, Z.-S.; Hung, F.-Y.; Chang, S.-J.; Hsieh, W.-K.; Chen, K.-J. Align Ag nanorods via oxidation reduction growth using RF-sputtering. *J. Nanomater.* **2012**, *2012*, 2. [CrossRef]
65. Lu, Y.; Zhang, C.-Y.; Zhang, D.-J.; Hao, R.; Hao, Y.-W.; Liu, Y.-Q. Fabrication of flower-like silver nanoparticles for surface-enhanced Raman scattering. *Chin. Chem. Lett.* **2016**, *27*, 689–692. [CrossRef]
66. Khorshidi, A.; Mardazad, N. Flower-like silver nanoparticles: An effective and recyclable catalyst for degradation of Rhodamine B with H_2O_2. *Res. Chem. Intermed.* **2016**, *42*, 7551–7558. [CrossRef]
67. Turkevich, J.; Stevenson, P.C.; Hillier, J. A study of the nucleation and growth processes in the synthesis of colloidal gold. *Discuss. Faraday Soc.* **1951**, *11*, 55–75. [CrossRef]
68. Lee, P.; Meisel, D. Adsorption and surface-enhanced Raman of dyes on silver and gold sols. *J. Phys. Chem.* **1982**, *86*, 3391–3395. [CrossRef]
69. Pillai, Z.S.; Kamat, P.V. What Factors control the size and shape of silver nanoparticles in the citrate ion reduction method? *J. Phys. Chem. B* **2004**, *108*, 945–951. [CrossRef]
70. Henglein, A.; Giersig, M. Formation of colloidal silver nanoparticles: Capping action of citrate. *J. Phys. Chem. B* **1999**, *103*, 9533–9539. [CrossRef]
71. Alqadi, M.; Noqtah, O.A.; Alzoubi, F.; Alzouby, J.; Aljarrah, K. pH effect on the aggregation of silver nanoparticles synthesized by chemical reduction. *Mater. Sci.-Pol.* **2014**, *32*, 107–111. [CrossRef]
72. Dong, X.; Ji, X.; Wu, H.; Zhao, L.; Li, J.; Yang, W. Shape control of silver nanoparticles by stepwise citrate reduction. *J. Phys. Chem. C* **2009**, *113*, 6573–6576. [CrossRef]

73. Qin, Y.; Ji, X.; Jing, J.; Liu, H.; Wu, H.; Yang, W. Size control over spherical silver nanoparticles by ascorbic acid reduction. *Colloids Surf. A Physicochem. Eng. Asp.* **2010**, *372*, 172–176. [CrossRef]

74. Jackson, J.B.; Halas, N.J. Silver nanoshells: Variations in morphologies and optical properties. *J. Phys. Chem. B* **2001**, *105*, 2743–2746. [CrossRef]

75. Jiang, Z.-J.; Liu, C.-Y. Seed-mediated growth technique for the preparation of a silver nanoshell on a silica sphere. *J. Phys. Chem. B* **2003**, *107*, 12411–12415. [CrossRef]

76. Halas, N. Playing with plasmons: Tuning the optical resonant properties of metallic nanoshells. *MRS Bull.* **2005**, *30*, 362–367. [CrossRef]

77. Murphy, C.J.; Sau, T.K.; Gole, A.M.; Orendorff, C.J.; Gao, J.; Gou, L.; Hunyadi, S.E.; Li, T. Anisotropic metal nanoparticles: Synthesis, assembly, and optical applications. *J. Phys. Chem. B* **2005**, *109*, 13857–13870. [CrossRef] [PubMed]

78. Li, N.; Zhao, P.; Astruc, D. Anisotropic gold nanoparticles: Synthesis, properties, applications, and toxicity. *Angew. Chem. Int. Ed.* **2014**, *53*, 1756–1789. [CrossRef] [PubMed]

79. Hirsch, L.R.; Gobin, A.M.; Lowery, A.R.; Tam, F.; Drezek, R.A.; Halas, N.J.; West, J.L. Metal nanoshells. *Ann Biomed. Eng.* **2006**, *34*, 15–22. [CrossRef] [PubMed]

80. Jakab, A.; Rosman, C.; Khalavka, Y.; Becker, J.; Trügler, A.; Hohenester, U.; Sönnichsen, C. Highly sensitive plasmonic silver nanorods. *ACS Nano* **2011**, *5*, 6880–6885. [CrossRef] [PubMed]

81. Jana, N.R.; Gearheart, L.; Murphy, C.J. Seed-mediated growth approach for shape-controlled synthesis of spheroidal and rod-like gold nanoparticles using a surfactant template. *Adv. Mater.* **2001**, *13*, 1389. [CrossRef]

82. Sun, Y.; Xia, Y. Shape-controlled synthesis of gold and silver nanoparticles. *Science* **2002**, *298*, 2176. [CrossRef]

83. Chang, S.; Chen, K.; Hua, Q.; Ma, Y.; Huang, W. Evidence for the growth mechanisms of silver nanocubes and nanowires. *J. Phys. Chem. C* **2011**, *115*, 7979–7986. [CrossRef]

84. Grzelczak, M.; Perez-Juste, J.; Mulvaney, P.; Liz-Marzan, L.M. Shape control in gold nanoparticle synthesis. *Chem. Soc. Rev.* **2008**, *37*, 1783–1791. [CrossRef]

85. Xia, Y.; Xiong, Y.; Lim, B.; Skrabalak, S.E. Shape-controlled synthesis of metal nanocrystals: Simple chemistry meets complex physics? *Angew. Chem. Int. Ed.* **2008**, *48*, 60–103. [CrossRef] [PubMed]

86. Orendorff, C.J.; Gearheart, L.; Jana, N.R.; Murphy, C.J. Aspect ratio dependence on surface enhanced Raman scattering using silver and gold nanorod substrates. *Phys. Chem. Chem. Phys.* **2006**, *8*, 165–170. [CrossRef] [PubMed]

87. Zaheer, Z.; Rafiuddin. Multi-branched flower-like silver nanoparticles: Preparation and characterization. *Colloids Surf. A Physicochem. Eng. Asp.* **2011**, *384*, 427–431. [CrossRef]

88. Murphy, C.J.; Jana, N.R. Controlling the aspect ratio of inorganic nanorods and nanowires. *Adv. Mater.* **2002**, *14*, 80–82. [CrossRef]

89. Lu, L.; Kobayashi, A.; Tawa, K.; Ozaki, Y. Silver nanoplates with special shapes: Controlled synthesis and their surface plasmon resonance and surface-enhanced Raman scattering properties. *Chem. Mater.* **2006**, *18*, 4894–4901. [CrossRef]

90. Pastoriza-Santos, I.; Liz-Marzán, L.M. Colloidal silver nanoplates. State of the art and future challenges. *J. Mater. Chem.* **2008**, *18*, 1724–1737. [CrossRef]

91. Potara, M.; Gabudean, A.-M.; Astilean, S. Solution-phase, dual LSPR-SERS plasmonic sensors of high sensitivity and stability based on chitosan-coated anisotropic silver nanoparticles. *J. Mater. Chem.* **2011**, *21*, 3625–3633. [CrossRef]

92. Liz-Marzán, L.M. Nanometals: Formation and color. *Mater. Today* **2004**, *7*, 26–31. [CrossRef]

93. Tan, T.; Tian, C.; Ren, Z.; Yang, J.; Chen, Y.; Sun, L.; Li, Z.; Wu, A.; Yin, J.; Fu, H. LSPR-dependent SERS performance of silver nanoplates with highly stable and broad tunable LSPRs prepared through an improved seed-mediated strategy. *Phys. Chem. Chem. Phys.* **2013**, *15*, 21034–21042. [CrossRef]

94. Wei, L.; Lu, J.; Xu, H.; Patel, A.; Chen, Z.-S.; Chen, G. Silver nanoparticles: Synthesis, properties, and therapeutic applications. *Drug Discov. Today* **2015**, *20*, 595–601. [CrossRef]

95. Millstone, J.E.; Hurst, S.J.; Métraux, G.S.; Cutler, J.I.; Mirkin, C.A. Colloidal gold and silver triangular nanoprisms. *Small* **2009**, *5*, 646–664. [CrossRef] [PubMed]

96. Zhou, W.; Ma, Y.; Yang, H.; Ding, Y.; Luo, X. A label-free biosensor based on silver nanoparticles array for clinical detection of serum p53 in head and neck squamous cell carcinoma. *Int. J. Nanomed.* **2011**, *6*, 381. [CrossRef] [PubMed]

97. Colson, P.; Henrist, C.; Cloots, R. Nanosphere lithography: A powerful method for the controlled manufacturing of nanomaterials. *J. Nanomater.* **2013**, *2013*, 21. [CrossRef]

98. Wiley, B.; Sun, Y.; Mayers, B.; Xia, Y. Shape-controlled synthesis of metal nanostructures: The case of silver. *Chem. Eur. J.* **2005**, *11*, 454–463. [CrossRef] [PubMed]

99. Im, S.H.; Lee, Y.T.; Wiley, B.; Xia, Y. Large-scale synthesis of silver nanocubes: The role of hcl in promoting cube perfection and monodispersity. *Angew. Chem.* **2005**, *117*, 2192–2195. [CrossRef]

100. Zeng, J.; Zheng, Y.; Rycenga, M.; Tao, J.; Li, Z.-Y.; Zhang, Q.; Zhu, Y.; Xia, Y. Controlling the shapes of silver nanocrystals with different capping agents. *J. Am. Chem. Soc.* **2010**, *132*, 8552–8553. [CrossRef] [PubMed]

101. Tao, A.; Sinsermsuksakul, P.; Yang, P. Polyhedral silver nanocrystals with distinct scattering signatures. *Angew. Chem. Int. Ed.* **2006**, *45*, 4597–4601. [CrossRef] [PubMed]

102. Siekkinen, A.R.; McLellan, J.M.; Chen, J.; Xia, Y. Rapid synthesis of small silver nanocubes by mediating polyol reduction with a trace amount of sodium sulfide or sodium hydrosulfide. *Chem. Phys. Lett.* **2006**, *432*, 491–496. [CrossRef]

103. Zhang, Q. (Ed.) *Nanotoxicity: Methods and Protocols*, 1st ed.; Humana Press, Springer: New York, NY, USA, 2019; Volume XII, 371p. [CrossRef]

104. Li, X.; Lenhart, J.J.; Walker, H.W. Dissolution-accompanied aggregation kinetics of silver nanoparticles. *Langmuir* **2010**, *26*, 16690–16698. [CrossRef]

105. Badawy, A.M.E.; Luxton, T.P.; Silva, R.G.; Scheckel, K.G.; Suidan, M.T.; Tolaymat, T.M. Impact of environmental conditions (pH, ionic strength, and electrolyte type) on the surface charge and aggregation of silver nanoparticles suspensions. *Environ. Sci. Technol.* **2010**, *44*, 1260–1266. [CrossRef]

106. Hotze, E.M.; Phenrat, T.; Lowry, G.V. Nanoparticle aggregation: Challenges to understanding transport and reactivity in the environment all rights reserved. No part of this periodical may be reproduced or transmitted in any form or by any means, electronic or mechanical, including photocopying, recording, or any information storage and retrieval system, without permission in writing from the publisher. *J. Environ. Qual.* **2010**, *39*, 1909–1924. [CrossRef] [PubMed]

107. Tejamaya, M.; Römer, I.; Merrifield, R.C.; Lead, J.R. Stability of citrate, PVP, and PEG coated silver nanoparticles in ecotoxicology media. *Environ. Sci. Technol.* **2012**, *46*, 7011–7017. [CrossRef] [PubMed]

108. Desireddy, A.; Conn, B.E.; Guo, J.; Yoon, B.; Barnett, R.N.; Monahan, B.M.; Kirschbaum, K.; Griffith, W.P.; Whetten, R.L.; Landman, U.; et al. Ultrastable silver nanoparticles. *Nature* **2013**, *501*, 399. [CrossRef] [PubMed]

109. Suman, T.; Rajasree, S.R.; Kanchana, A.; Elizabeth, S.B. Biosynthesis, characterization and cytotoxic effect of plant mediated silver nanoparticles using Morinda citrifolia root extract. *Colloids Surf. B Biointerfaces* **2013**, *106*, 74–78. [CrossRef] [PubMed]

110. Kvitek, L.; Panáček, A.; Soukupova, J.; Kolář, M.; Večeřová, R.; Prucek, R.; Holecova, M.; Zbořil, R. Effect of surfactants and polymers on stability and antibacterial activity of silver nanoparticles (NPs). *J. Phys. Chem. C* **2008**, *112*, 5825–5834. [CrossRef]

111. Wiley, B.; Sun, Y.; Xia, Y. Synthesis of silver nanostructures with controlled shapes and properties. *Acc. Chem. Res.* **2007**, *40*, 1067–1076. [CrossRef]

112. Hardikar, V.V.; Matijević, E. Coating of nanosize silver particles with silica. *J. Colloid Int. Sci.* **2000**, *221*, 133–136. [CrossRef] [PubMed]

113. Brandon, M.P.; Ledwith, D.M.; Kelly, J.M. Preparation of saline-stable, silica-coated triangular silver nanoplates of use for optical sensing. *J. Colloid Int. Sci.* **2014**, *415*, 77–84. [CrossRef]

114. Levard, C.; Hotze, E.M.; Lowry, G.V.; Brown, G.E., Jr. Environmental transformations of silver nanoparticles: Impact on stability and toxicity. *Environ. Sci. Technol.* **2012**, *46*, 6900–6914. [CrossRef]

115. Sharma, V.K.; Siskova, K.M.; Zboril, R.; Gardea-Torresdey, J.L. Organic-coated silver nanoparticles in biological and environmental conditions: Fate, stability and toxicity. *Adv. Colloid Interface Sci.* **2014**, *204*, 15–34. [CrossRef]

116. Delay, M.; Dolt, T.; Woellhaf, A.; Sembritzki, R.; Frimmel, F.H. Interactions and stability of silver nanoparticles in the aqueous phase: Influence of natural organic matter (NOM) and ionic strength. *J. Chromatogr. A* **2011**, *1218*, 4206–4212. [CrossRef] [PubMed]

117. Gunsolus, I.L.; Mousavi, M.P.; Hussein, K.; Bühlmann, P.; Haynes, C.L. Effects of humic and fulvic acids on silver nanoparticle stability, dissolution, and toxicity. *Environ. Sci. Technol.* **2015**, *49*, 8078–8086. [CrossRef] [PubMed]

118. Farhadi, K.; Forough, M.; Molaei, R.; Hajizadeh, S.; Rafipour, A. Highly selective Hg^{2+} colorimetric sensor using green synthesized and unmodified silver nanoparticles. *Sens. Actuators B Chem.* **2012**, *161*, 880–885. [CrossRef]
119. Battocchio, C.; Meneghini, C.; Fratoddi, I.; Venditti, I.; Russo, M.V.; Aquilanti, G.; Maurizio, C.; Bondino, F.; Matassa, R.; Rossi, M.; et al. Silver nanoparticles stabilized with thiols: A close look at the local chemistry and chemical structure. *J. Phys. Chem. C* **2012**, *116*, 19571–19578. [CrossRef]
120. Cheng, X.; Liu, M.; Zhang, A.; Hu, S.; Song, C.; Zhang, G.; Guo, X. Size-controlled silver nanoparticles stabilized on thiol-functionalized MIL-53(Al) frameworks. *Nanoscale* **2015**, *7*, 9738–9745. [CrossRef] [PubMed]
121. Brust, M.; Walker, M.; Bethell, D.; Schiffrin, D.J.; Whyman, R. Synthesis of thiol-derivatised gold nanoparticles in a two-phase Liquid-Liquid system. *J. Chem. Soc. Chem. Commun.* **1994**, *7*. [CrossRef]
122. Popa, M.; Pradell, T.; Crespo, D.; Calderón-Moreno, J.M. Stable silver colloidal dispersions using short chain polyethylene glycol. *Colloids Surf. A Physicochem. Eng. Asp.* **2007**, *303*, 184–190. [CrossRef]
123. Papa, A.-L.; Boudon, J.; Bellat, V.; Loiseau, A.; Bisht, H.; Sallem, F.; Chassagnon, R.; Berard, V.; Millot, N. Dispersion of titanate nanotubes for nanomedicine: Comparison of PEI and PEG nanohybrids. *Dalton Trans.* **2015**, *44*, 739–746. [CrossRef]
124. Mosqueira, V.C.F.; Legrand, P.; Morgat, J.-L.; Vert, M.; Mysiakine, E.; Gref, R.; Devissaguet, J.-P.; Barratt, G. Biodistribution of long-circulating PEG-grafted nanocapsules in mice: Effects of PEG chain length and density. *Pharm. Res.* **2001**, *18*, 1411–1419. [CrossRef]
125. Cruje, C.; Chithrani, B. Polyethylene glycol density and length affects nanoparticle uptake by cancer cells. *J. Nanomed. Res.* **2014**, *1*, 00006. [CrossRef]
126. Maurizi, L.; Papa, A.-L.; Dumont, L.; Bouyer, F.; Walker, P.; Vandroux, D.; Millot, N. Influence of surface charge and polymer coating on internalization and biodistribution of polyethylene glycol-modified iron oxide nanoparticles. *J. Biomed. Nanotechnol.* **2015**, *11*, 126–136. [CrossRef] [PubMed]
127. Karakoti, A.S.; Das, S.; Thevuthasan, S.; Seal, S. PEGylated inorganic nanoparticles. *Angew. Chem. Int. Ed.* **2011**, *50*, 1980–1994. [CrossRef] [PubMed]
128. Molineux, G. Pegylation: Engineering improved biopharmaceuticals for oncology. *Pharmacother. J. Hum. Pharmacol. Drug Ther.* **2003**, *23*, 3S–8S. [CrossRef]
129. Bastos, V.; Ferreira de Oliveira, J.M.P.; Brown, D.; Jonhston, H.; Malheiro, E.; Daniel-da-Silva, A.L.; Duarte, I.F.; Santos, C.; Oliveira, H. The influence of citrate or PEG coating on silver nanoparticle toxicity to a human keratinocyte cell line. *Toxicol. Lett.* **2016**, *249*, 29–41. [CrossRef] [PubMed]
130. Shameli, K.; Bin Ahmad, M.; Jazayeri, S.D.; Sedaghat, S.; Shabanzadeh, P.; Jahangirian, H.; Mahdavi, M.; Abdollahi, Y. Synthesis and characterization of polyethylene glycol mediated silver nanoparticles by the green method. *Int. J. Mol. Sci.* **2012**, *13*, 6639. [CrossRef] [PubMed]
131. Sun, T.; Zhang, Y.S.; Pang, B.; Hyun, D.C.; Yang, M.; Xia, Y. Engineered nanoparticles for drug delivery in cancer therapy. *Angew. Chem. Int. Ed.* **2014**, *53*, 12320–12364. [CrossRef]
132. Sallem, F.; Boudon, J.; Heintz, O.; Séverin, I.; Megriche, A.; Millot, N. Synthesis and characterization of chitosan-coated titanate nanotubes: Towards a new safe nanocarrier. *Dalton Trans.* **2017**. [CrossRef]
133. Jiang, H.; Chen, Z.; Cao, H.; Huang, Y. Peroxidase-like activity of chitosan stabilized silver nanoparticles for visual and colorimetric detection of glucose. *Analyst* **2012**, *137*, 5560–5564. [CrossRef]
134. Guerrero-Martínez, A.; Pérez-Juste, J.; Liz-Marzán, L.M. Recent progress on silica coating of nanoparticles and related nanomaterials. *Adv. Mater.* **2010**, *22*, 1182–1195. [CrossRef]
135. Liu, S.; Han, M.-Y. Silica-coated metal nanoparticles. *Chem. Asian J.* **2010**, *5*, 36–45. [CrossRef]
136. Boujday, S.; Parikh, A.N.; Liedberg, B.; Song, H. Functionalisation of gold nanoparticles. In *Gold Nanoparticles for Physics, Chemistry and Biology*; Louis, C., Pluchery, O., Eds.; World Scientific: London, UK, 2017; pp. 201–227. [CrossRef]
137. Aissaoui, N.; Bergaoui, L.; Landoulsi, J.; Lambert, J.-F.; Boujday, S. Silane layers on silicon surfaces: Mechanism of interaction, stability, and influence on protein adsorption. *Langmuir* **2012**, *28*, 656–665. [CrossRef] [PubMed]
138. Hanske, C.; Sanz-Ortiz, M.N.; Liz-Marzán, L.M. Silica-coated plasmonic metal nanoparticles in action. *Adv. Mater.* **2018**, *30*, 1707003. [CrossRef] [PubMed]
139. Liz-Marzán, L.M.; Giersig, M.; Mulvaney, P. Synthesis of nanosized gold−silica core−shell particles. *Langmuir* **1996**, *12*, 4329–4335. [CrossRef]
140. Kobayashi, Y.; Katakami, H.; Mine, E.; Nagao, D.; Konno, M.; Liz-Marzán, L.M. Silica coating of silver nanoparticles using a modified Stöber method. *J. Colloid Int. Sci.* **2005**, *283*, 392–396. [CrossRef] [PubMed]

141. Liu, S.; Zhang, Z.; Han, M. Gram-scale synthesis and biofunctionalization of silica-coated silver nanoparticles for fast colorimetric DNA detection. *Anal. Chem.* **2005**, *77*, 2595–2600. [CrossRef]

142. Sotiriou, G.A.; Sannomiya, T.; Teleki, A.; Krumeich, F.; Vörös, J.; Pratsinis, S.E. Non-toxic dry-coated nanosilver for plasmonic biosensors. *Adv. Funct. Mater.* **2010**, *20*, 4250–4257. [CrossRef] [PubMed]

143. Lismont, M.; Dreesen, L. Comparative study of Ag and Au nanoparticles biosensors based on surface plasmon resonance phenomenon. *Mater. Sci. Eng. C* **2012**, *32*, 1437–1442. [CrossRef] [PubMed]

144. Link, S.; Wang, Z.L.; El-Sayed, M.A. Alloy formation of gold–silver nanoparticles and the dependence of the plasmon absorption on their composition. *J. Phys. Chem. B* **1999**, *103*, 3529–3533. [CrossRef]

145. Navas, M.P.; Soni, R.K. Laser generated Ag and Ag–Au composite nanoparticles for refractive index sensor. *Appl. Phys. A* **2014**, *116*, 879–886. [CrossRef]

146. Jia, K.; Khaywah, M.Y.; Li, Y.; Bijeon, J.L.; Adam, P.M.; Déturche, R.G.; Guelorget, B.; François, M.; Louarn, G.; Ionescu, R.E. Strong improvements of localized surface plasmon resonance sensitivity by using Au/Ag bimetallic nanostructures modified with polydopamine films. *ACS Appl. Mater. Int.* **2013**, *6*, 219–227. [CrossRef]

147. Heinz, M.; Srabionyan, V.V.; Avakyan, L.A.; Bugaev, A.L.; Skidanenko, A.V.; Kaptelinin, S.Y.; Ihlemann, J.; Meinertz, J.; Patzig, C.; Dubiel, M.; et al. Formation of bimetallic gold-silver nanoparticles in glass by UV laser irradiation. *J. Alloy. Compd.* **2018**, *767*, 1253–1263. [CrossRef]

148. Wu, W.; Njoki, P.N.; Han, H.; Zhao, H.; Schiff, E.A.; Lutz, P.S.; Solomon, L.; Matthews, S.; Maye, M.M. Processing core/alloy/shell nanoparticles: Tunable optical properties and evidence for self-limiting alloy growth. *J. Phys. Chem. C* **2011**, *115*, 9933–9942. [CrossRef]

149. Sancho-Parramon, J.; Janicki, V.; Lončarić, M.; Zorc, H.; Dubček, P.; Bernstorff, S. Optical and structural properties of Au-Ag islands films for plasmonic applications. *Appl. Phys. A* **2011**, *103*, 745–748. [CrossRef]

150. Michieli, N.; Kalinic, B.; Scian, C.; Cesca, T.; Mattei, G. Optimal geometric parameters of ordered arrays of nanoprisms for enhanced sensitivity in localized plasmon based sensors. *Biosens. Bioelectron.* **2015**, *65*, 346–353. [CrossRef] [PubMed]

151. Gao, C.; Lu, Z.; Liu, Y.; Zhang, Q.; Chi, M.; Cheng, Q.; Yin, Y. Highly stable silver nanoplates for surface plasmon resonance biosensing. *Angew. Chem. Int. Ed.* **2012**, *51*, 5629–5633. [CrossRef] [PubMed]

152. Dong, P.; Lin, Y.; Deng, J.; Di, J. Ultrathin gold-shell coated silver nanoparticles onto a glass platform for improvement of plasmonic sensors. *ACS Appl. Mater. Int.* **2013**, *5*, 2392–2399. [CrossRef] [PubMed]

153. Yang, X.; Ren, Y.; Gao, Z. Silver/gold core-shell nanoprism-based plasmonic nanoprobes for highly sensitive and selective detection of hydrogen sulfide. *Chem. Eur. J.* **2014**, *21*, 988–992. [CrossRef] [PubMed]

154. Ma, Y.; Zhou, J.; Zou, W.; Jia, Z.; Petti, L.; Mormile, P. Localized surface plasmon resonance and surface enhanced Raman scattering responses of Au@Ag core-shell nanorods with different thickness of Ag shell. *Nanosci. Nanotechnol.* **2013**, *13*, 1–6. [CrossRef]

155. Zhu, J.; Zhang, F.; Li, J.-J.; Zhao, J.-W. Optimization of the refractive index plasmonic sensing of gold nanorods by non-uniform silver coating. *Sens. Actuators B Chem.* **2013**, *183*, 556–564. [CrossRef]

156. Sun, L.; Li, Q.; Tang, W.; Di, J.; Wu, Y. The use of gold-silver core-shell nanorods self-assembled on a glass substrate can substantially improve the performance of plasmonic affinity biosensors. *Microchimica Acta* **2014**, *181*, 1991–1997. [CrossRef]

157. Dong, P.; Wu, Y.; Guo, W.; Di, J. Plasmonic biosensor based on triangular Au/Ag and Au/Ag/Au core/shell nanoprisms onto indium tin oxide glass. *Plasmonics* **2013**, *8*, 1577–1583. [CrossRef]

158. Singh, A.V.; Bandgar, B.M.; Kasture, M.; Prasad, B.L.V.; Sastry, M. Synthesis of gold, silver and their alloy nanoparticles using bovine serum albumin as foaming and stabilizing agent. *J. Mater. Chem.* **2005**, *15*, 5115–5121. [CrossRef]

159. Qi, W.; Lee, S. Phase stability, melting, and alloy formation of Au-Ag bimetallic nanoparticles. *J. Phys. Chem. C* **2010**, *114*, 9580–9587. [CrossRef]

160. Mallin, M.P.; Murphy, C.J. Solution-phase synthesis of Sub-10 nm Au-Ag alloy nanoparticles. *Nano Lett.* **2002**, *2*, 1235–1237. [CrossRef]

161. Liz-Marzán, L.M. Tailoring surface plasmons through the morphology and assembly of metal nanoparticles. *Langmuir* **2006**, *22*, 32–41. [CrossRef]

162. Pérez-Juste, J.; Pastoriza-Santos, I.; Liz-Marzán, L.M.; Mulvaney, P. Gold nanorods: Synthesis, characterization and applications. *Coord. Chem. Rev.* **2005**, *249*, 1870–1901. [CrossRef]

163. Zhang, L.; Chen, P.; Loiseau, A.; Brouri, D.; Casale, S.; Salmain, M.; Boujday, S.; Liedberg, B. Spatially controlled reduction and growth of silver in hollow gold nanoshell particles. *J. Phys. Chem. C* **2019**, *123*, 10614–10621. [CrossRef]

164. Rodríguez-González, B.; Burrows, A.; Watanabe, M.; Kiely, C.J.; Liz Marzán, L.M. Multishell bimetallic AuAg nanoparticles: Synthesis, structure and optical properties. *J. Mater. Chem.* **2005**, *15*, 1755–1759. [CrossRef]

165. Chen, Y.; Wu, H.; Li, Z.; Wang, P.; Yang, L.; Fang, Y. The study of surface plasmon in Au/Ag core/shell compound nanoparticles. *Plasmonics* **2012**, *7*, 509–513. [CrossRef]

166. Zhu, J. Surface plasmon resonance from bimetallic interface in Au-Ag core-shell structure nanowires. *Nanoscale Res. Lett.* **2009**, *4*, 977. [CrossRef]

167. Jiang, R.; Chen, H.; Shao, L.; Li, Q.; Wang, J. Unraveling the evolution and nature of the plasmons in (Au core)-(Ag shell) nanorods. *Adv. Mater.* **2012**, *24*, OP200–OP207. [CrossRef] [PubMed]

168. Becker, J.; Zins, I.; Jakab, A.; Khalavka, Y.; Schubert, O.; Sönnichsen, C. Plasmonic focusing reduces ensemble linewidth of silver-coated gold nanorods. *Nano Lett.* **2008**, *8*, 1719–1723. [CrossRef] [PubMed]

169. Verma, S.; Rao, B.T.; Detty, A.P.; Ganesan, V.; Phase, D.M.; Rai, S.K.; Bose, A.; Joshi, S.C.; Kukreja, L.M. Surface plasmon resonances of Ag-Au alloy nanoparticle films grown by sequential pulsed laser deposition at different compositions and temperatures. *J. Appl. Phys.* **2015**, *117*, 133105. [CrossRef]

170. Zhu, J.; Li, X.; Li, J.-J.; Zhao, J.-W. Enlarge the biologic coating-induced absorbance enhancement of Au-Ag bimetallic nanoshells by tuning the metal composition. *Spectrochim. Acta Part A Mol. Biomol. Spectrosc.* **2018**, *189*, 571–577. [CrossRef] [PubMed]

171. Jana, D.; Lehnhoff, E.; Bruzas, I.; Robinson, J.; Lum, W.; Sagle, L. Tunable Au-Ag nanobowl arrays for size-selective plasmonic biosensing. *Analyst* **2016**, *141*, 4870–4878. [CrossRef] [PubMed]

172. Skrabalak, S.E.; Chen, J.; Sun, Y.; Lu, X.; Au, L.; Cobley, C.M.; Xia, Y. Gold Nanocages: Synthesis, properties, and applications. *Acc. Chem. Res.* **2008**, *41*, 1587–1595. [CrossRef] [PubMed]

173. Yan, Z.; Wu, Y.; Di, J. Formation of substrate-based gold nanocage chains through dealloying with nitric acid. *Beilstein J. Nanotechnol.* **2015**, *6*, 1362. [CrossRef] [PubMed]

174. Goodman, A.M.; Cao, Y.; Urban, C.; Neumann, O.; Ayala-Orozco, C.; Knight, M.W.; Joshi, A.; Nordlander, P.; Halas, N.J. The surprising in vivo instability of near-IR-absorbing hollow Au-Ag nanoshells. *ACS Nano* **2014**, *8*, 3222–3231. [CrossRef] [PubMed]

175. Russo, L.; Merkoçi, F.; Patarroyo, J.; Piella, J.; Merkoçi, A.; Bastús, N.G.; Puntes, V. Time- and size-resolved plasmonic evolution with nm resolution of galvanic replacement reaction in auag nanoshells synthesis. *Chem. Mater.* **2018**, *30*, 5098–5107. [CrossRef]

176. Prevo, B.G.; Esakoff, S.A.; Mikhailovsky, A.; Zasadzinski, J.A. Scalable routes to gold nanoshells with tunable sizes and response to near-infrared pulsed-laser irradiation. *Small* **2008**, *4*, 1183–1195. [CrossRef]

177. Hall, W.P.; Ngatia, S.N.; van Duyne, R.P. LSPR Biosensor signal enhancement using nanoparticle-antibody conjugates. *J. Phys. Chem. C* **2011**, *115*, 1410–1414. [CrossRef] [PubMed]

178. Ortega-Mendoza, G.J.; Padilla-Vivanco, A.; Toxqui-Quitl, C.; Zaca-Morán, P.; Villegas-Hernández, D.; Chávez, F. Optical fiber sensor based on localized surface plasmon resonance using silver nanoparticles photodeposited on the optical fiber end. *Sensors* **2014**, *14*, 18701–18710. [CrossRef] [PubMed]

179. Chen, J.; Shi, S.; Su, R.; Qi, W.; Huang, R.; Wang, M.; Wang, L.; He, Z. Optimization and application of reflective lspr optical fiber biosensors based on silver nanoparticles. *Sensors* **2015**, *15*, 12205–12217. [CrossRef] [PubMed]

180. Sagle, L.B.; Ruvuna, L.K.; Ruemmele, J.A.; van Duyne, R.P. Advances in localized surface plasmon resonance spectroscopy biosensing. *Nanomedicine* **2011**, *6*, 1447–1462. [CrossRef] [PubMed]

181. Samal, A.K.; Polavarapu, L.; Rodal-Cedeira, S.; Liz-Marzán, L.M.; Pérez-Juste, J.; Pastoriza-Santos, I. Size tunable Au@Ag core-shell nanoparticles: Synthesis and surface-enhanced raman scattering properties. *Langmuir* **2013**, *29*, 15076–15082. [CrossRef]

182. Vilela, D.; González, M.C.; Escarpa, A. Sensing colorimetric approaches based on gold and silver nanoparticles aggregation: Chemical creativity behind the assay: A review. *Anal. Chim. Acta* **2012**, *751*, 24–43. [CrossRef]

183. Han, C.; Li, H. Visual detection of melamine in infant formula at 0.1 ppm level based on silver nanoparticles. *Analyst* **2010**, *135*, 583–588. [CrossRef]

184. Ma, Y.; Niu, H.; Zhang, X.; Cai, Y. One-step synthesis of silver/dopamine nanoparticles and visual detection of melamine in raw milk. *Analyst* **2011**, *136*, 4192–4196. [CrossRef]

185. Li, H.; Li, F.; Han, C.; Cui, Z.; Xie, G.; Zhang, A. Highly sensitive and selective tryptophan colorimetric sensor based on 4,4-bipyridine-functionalized silver nanoparticles. *Sens. Actuators B Chem.* **2010**, *145*, 194–199. [CrossRef]
186. Xiong, D.; Chen, M.; Li, H. Synthesis of para-sulfonatocalix[4]arene-modified silver nanoparticles as colorimetric histidine probes. *Chem. Commun.* **2008**, 880–882. [CrossRef]
187. Xiong, D.; Li, H. Colorimetric detection of pesticides based on calixarene modified silver nanoparticles in water. *Nanotechnology* **2008**, *19*, 465502. [CrossRef] [PubMed]
188. Guo, Y.; Wu, J.; Li, J.; Ju, H. A plasmonic colorimetric strategy for biosensing through enzyme guided growth of silver nanoparticles on gold nanostars. *Biosens. Bioelectron.* **2016**, *78*, 267–273. [CrossRef] [PubMed]
189. Mao, K.; Yang, Z.; Li, J.; Zhou, X.; Li, X.; Hu, J. A novel colorimetric biosensor based on non-aggregated Au@Ag core-shell nanoparticles for methamphetamine and cocaine detection. *Talanta* **2017**, *175*, 338–346. [CrossRef] [PubMed]
190. Aslan, K.; Lakowicz, J.R.; Szmacinski, H.; Geddes, C.D. Metal-enhanced fluorescence solution-based sensing platform. *J. Fluoresc.* **2004**, *14*, 677–679. [CrossRef]
191. Aslan, K.; Gryczynski, I.; Malicka, J.; Matveeva, E.; Lakowicz, J.R.; Geddes, C.D. Metal-enhanced fluorescence: An emerging tool in biotechnology. *Curr. Opin. Biotechnol.* **2005**, *16*, 55–62. [CrossRef] [PubMed]
192. Malicka, J.; Gryczynski, I.; Lakowicz, J.R. DNA hybridization assays using metal-enhanced fluorescence. *Biochem. Biophys. Res. Commun.* **2003**, *306*, 213–218. [CrossRef]
193. Zhang, J.; Malicka, J.; Gryczynski, I.; Lakowicz, J.R. Oligonucleotide-displaced organic monolayer-protected silver nanoparticles and enhanced luminescence of their salted aggregates. *Anal. Biochem.* **2004**, *330*, 81–86. [CrossRef] [PubMed]
194. Aslan, K.; Wu, M.; Lakowicz, J.R.; Geddes, C.D. Fluorescent core-;shell Ag@SiO$_2$ nanocomposites for metal-enhanced fluorescence and single nanoparticle sensing platforms. *J. Am. Chem. Soc.* **2007**, *129*, 1524–1525. [CrossRef] [PubMed]
195. Yun, B.J.; Kwon, J.E.; Lee, K.; Koh, W.-G. Highly sensitive metal-enhanced fluorescence biosensor prepared on electrospun fibers decorated with silica-coated silver nanoparticles. *Sens. Actuators B Chem.* **2019**, *284*, 140–147. [CrossRef]
196. Xia, Y.; Ye, J.; Tan, K.; Wang, J.; Yang, G. Colorimetric visualization of glucose at the submicromole level in serum by a homogenous silver nanoprism–glucose oxidase system. *Anal. Chem.* **2013**, *85*, 6241–6247. [CrossRef]
197. Endo, T.; Ikeda, R.; Yanagida, Y.; Hatsuzawa, T. Stimuli-responsive hydrogel-silver nanoparticles composite for development of localized surface plasmon resonance-based optical biosensor. *Anal. Chim. Acta* **2008**, *611*, 205–211. [CrossRef] [PubMed]
198. He, H.; Xu, X.; Wu, H.; Jin, Y. Enzymatic plasmonic engineering of Ag/Au bimetallic nanoshells and their use for sensitive optical glucose sensing. *Adv. Mater.* **2012**, *24*, 1736–1740. [CrossRef] [PubMed]
199. Zhang, X.; Wei, M.; Lv, B.; Liu, Y.; Liu, X.; Wei, W. Sensitive colorimetric detection of glucose and cholesterol by using Au@Ag core-shell nanoparticles. *RSC Adv.* **2016**, *6*, 35001–35007. [CrossRef]

Review

Bio-Recognition in Spectroscopy-Based Biosensors for *Heavy Metals*-Water and Waterborne Contamination Analysis

Alessandra Aloisi [1,†], Antonio Della Torre [1,†], Angelantonio De Benedetto [2] and Rosaria Rinaldi [1,2,3,*]

[1]   Institute for Microelectronics and Microsystems (IMM), CNR, Via Monteroni, 73100 Lecce, Italy
[2]   Mathematics and Physics "E. De Giorgi" Department, University of Salento, Via Monteroni, 73100 Lecce, Italy
[3]   ISUFI, University of Salento, Via Monteroni, 73100 Lecce, Italy
*   Correspondence: ross.rinaldi@unisalento.it
†   These authors contributed equally to this work.

Received: 12 June 2019; Accepted: 25 July 2019; Published: 30 July 2019

Abstract: Microsystems and biomolecules integration as well multiplexing determinations are key aspects of sensing devices in the field of heavy metal contamination monitoring. The present review collects the most relevant information about optical biosensors development in the last decade. Focus is put on analytical characteristics and applications that are dependent on: (i) Signal transduction method (luminescence, colorimetry, evanescent wave (EW), surface-enhanced Raman spectroscopy (SERS), Förster resonance energy transfer (FRET), surface plasmon resonance (SPR); (ii) biorecognition molecules employed (proteins, nucleic acids, aptamers, and enzymes). The biosensing systems applied (or applicable) to water and milk samples will be considered for a comparative analysis, with an emphasis on water as the primary source of possible contamination along the food chain.

Keywords: water pollution; environmental water; drinking water; milk; heavy metal ions; biosensor; detection limits; optical spectroscopy; proteins; functional nucleic acids

---

1. Introduction

Biosensors are currently valid tools, other than laboratory analytical instrumentation, for monitoring the quality of natural water (e.g., in the food production chain) [1]. Biosensors are not meant to take over standard analytical methods, but, when optimal features of a sensing device are met, they offer remarkable advantages over conventional techniques. Overall, in certain conditions, their promptness and low-cost manufacturing make them useful tools to analyze many samples for primary warnings. As defined by the International Union of Pure and Applied Chemistry (IUPAC), "a biosensor is an integrated receptor ± transducer device, capable of providing selective analytical information using a biological recognition element" [2]. Optical biosensors are a group of sensors in which (i) the transducer senses optical fluctuations in the input light resultant from bioreceptor—target interaction, and (ii) the amplitude of these changes hinge on the concentration of the analyte [1].

Even in very small amounts, several metal ions may have important effects on health state, as they are hardly degradable but easily accumulated in the body through the diet [3]. Metal ions are generally not essential nutrients; conversely, they could be damaging to all living species [4].

* Widely indicated as "heavy metals" (HMs), in a technical report of 2002, the author concluded: "The term *heavy metal* has never been defined by any authoritative body such as IUPAC. No relationship can be found between density and any of the various physicochemical concepts that have been used to define *heavy metals* and the toxicity attributed to *heavy metals* ... Understanding bioavailability

is the key to assessment of the potential toxicity... It depends on biological parameters and on the physicochemical properties of metallic elements, their ions, and their compounds. These in turn depend upon the atomic structure of the metallic elements, systematically described by the periodic table" [5].

In the last twenty years, with the aim to quantify trace amounts of such possible contaminants, environmental monitoring has generated a need for innovative and improved approaches that have ever-increasing sensitivity and selectivity, as described in a recent review paper on various analytical techniques-based biosensors [6]. The introduction of biosensors has brought in new and promising approaches, but with still limited application in the environmental field if compared with the biomedical one, where most efforts have converged in the past years.

Much research is still needed before biosensors consolidate as a recognized analytical strategy with respect to environmental and food trace contaminant detection.

In this direction, the integration of nanomaterials and functional biological molecules is part of a new era in the optical biosensor area. Actually, nano-structured materials unveil distinctive size- and shape-dependent physicochemical properties, showing a number of possible interactions [7] with the biorecognition component, which may act as a reaction catalyst, or may be in equilibrium with macromolecules present in their natural biological settings or isolated and engineered [2]. Essentially, while the sensor sensitivity is influenced by the selected transducer component, the bioreceptor is responsible for the specificity [8]. Many biosensing elements that can be coupled to different transducers are now available for HM detection (Figure 1).

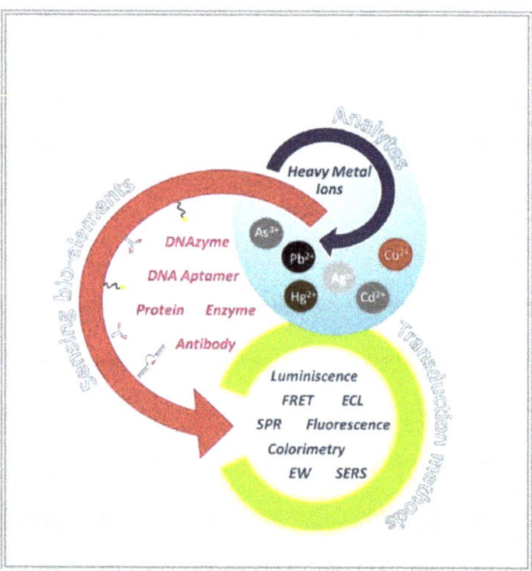

Figure 1. Optical biosensor scheme strategies for heavy metal (HM) ion detection in a water/milk "drop". Transduction methods and bioreceptor classes synergistically employed for the development of recently published devices.

In general, depending on the specific mechanism of the bioreceptor component, five groups can be identified: (i) DNA-based metal biosensor, (ii) antibodies, (iii) proteins, (iv) cellular structures or whole cells, (v) biomimetic receptors (gene-engineered molecules, molecularly imprinted polymers [9], and molecularly imprinted membranes [10], mimicking a natural bioreceptor. Most of them, natural and synthetic, are exhaustively described in recent review papers (concerning the interaction of metal ions with DNAs, peptides and enzymes, whole cells, as well as ionophores and small molecules) [8,11–15].

From a functional point of view, optical biosensors can be further categorized as: (i) probing biosensors: Entailing sensors based on target and recognition element affinity interaction; (ii) reacting biosensors: Where the optical responses relies on chemical processes [16]. Concerning the biomolecular probes, the most widely exploited can be collected into two macro groups: proteins and nucleic acids.

The specific affinities of these two families of molecules for HM ions are briefly introduced below, before entering the focal topic of this paper.

With regard to metal binding proteins, phytochelatins or metallothionein, metal ligands found in plants, are usually exploited on the surface of the transducer, where protein–metal interactions occur through the formation of a complex [17,18]. Functional proteins with enzymatic activity (purified or directly in a microorganism) catalyze specific chemical reactions also in the presence of metal ions. The mechanisms of action of these elements embrace: (a) Transformation of the analyte into a sensor-detectable product, (b) detection of an analyte behaving as negative or positive enzyme activity modulator, or (iii) appraisal of enzyme properties deviations upon interaction with the analyte [19,20].

Metal ions affinity for amino acid side chains (with sulfur, nitrogen, and oxygen atoms) and the occurrence of such amino acids in antibody-determining regions are expected to influence the ability of antibodies to strongly bind to metal–chelate complexes [21–24].

On a parallel route, functional nucleic acids (FNAs) represent molecules whose usefulness is further than that of encoding genetic information [25], and whose chemical structure is suitable for metal recognition. Two active structures have been developed for this purpose, working as either direct metal binding or metal-assisted deoxyribonucleic/ribonucleic acid catalyst. Definitely, aptamers, metal ion-specific DNA, guanine (G)-rich oligonucleotides, and DNA-based enzymes (DNAzymes) are the most widely reported [26]. In brief, aptamers are able to effectively bind basically any molecule of choice; they consist of artificial short single-stranded (ss) nucleic acid sequences or peptide molecules identified by combinatorial selection, through the Systematic evolution of ligands by exponential enrichment (SELEX) methodology [27,28] that, upon binding to targets, can fold into specific secondary and tertiary structures [29,30].

Basically, DNA and metal ions may interact in three different ways: (i) By HM ions-based exchanging of hydrogen atoms of the Watson–Crick base pairs; (ii) by reversible binding of HM ions with DNA; (iii) forming kinetically inert complexes by persistent crosslinking of DNA with HM ions [31]. The metal ion-specific DNAs are those sequences most commonly rich in thymine (T) or cytosine (C), with great selectivity for metal ions, which promote robust metal-base complex formation—specifically forming $T–Hg^{2+}–T$ [32] and $C–Ag^+–C$ mismatch [33]. G-rich DNAs are G-rich strands with a tendency to self-associate into non-canonical secondary structures named G-quadruplexes (G4) [34]. On these cations coordination-induced structure/property changes, a number of strategies have been proposed for the detection of Pb^{2+}, Hg^{2+}, Ba^{2+}, Ag^{2+}, K^+ [26].

A different class is then represented by nucleic acid enzymes (Ribozymes and DNAzymes). These are molecules found in nature like catalytic RNA or in vitro selected DNA sequences, displaying specific strong metal-dependent activity and structure recognition capability, bypassing the need for metal immobilization [25,26].

Remarkably, the choice of a suitable biological element and transduction module makes the biosensor sensitive and analyte specific, thus efficient for toxicological studies. Portable biosensors also make in-situ analysis possible, facilitating real-time monitoring [35].

In this context, the present work aims to review the sensitivity of HM-dedicated optical biosensor systems published in the last decade. Several biosensors relevant for water sample or liquid food monitoring are here described, although only those showing HM ion detection in real and complex matrices are compared, as reported in Table 1.

All methods are listed in order of the prevalence of published biosensors for HM sensing in water or milk matrix, be it a real or laboratory-built aqueous sample. Analytical techniques here presented include luminescence, colorimetry, evanescent wave, surface-enhanced Raman spectroscopy, Förster resonance energy transfer, and surface plasmon resonance. As the core purpose of this review is to recognize which method displays the maximum stated sensitivity—for the selected HM ion, focusing on the biosensing element employed—additional focused tables (Tables 2 and 3) have been worked out and introduced later in the text.

2. Biosensing Methods

In order to introduce a brief summary of what the reader will encounter during this paragraph, in Figure 2, the HM ion optical biosensor distribution is plotted with respect to the recognition element used, in the frame of the same transduction method, as already classified in Table 1.

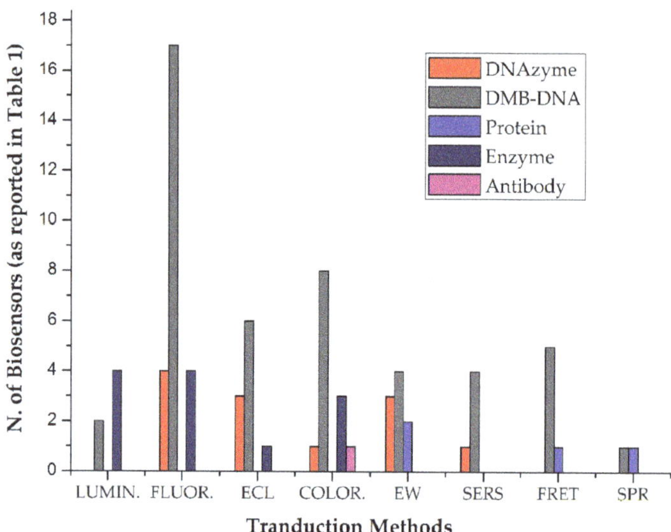

Figure 2. Distribution of biorecognition elements exploited in recently reported sensors for HM detection in real samples, as classified in Table 1.

With regards to the already mentioned classes of molecules, some considerations have emerged: (1) FNAs are the most employed; (2) direct metal binding DNA sequences (DMB-DNA) subclass, comprising aptamer, metal ion-specific DNA, and G-rich oligonucleotide, occupies a wide portion in the described FNAs-operating sensors; (3) proteins are the least employed, and (4) catalytic active protein-based sensors have been proposed more than those exploiting a non-catalytic protein, or a specific antibody.

In the next subparagraphs, the newly developed biosensors based on these recognition elements will be described, and with regard to the exploited biosensing mechanism, the more representative strategies will be showed in summary figures (Figures 3–10).

2.1. Luminescence

Luminescence concerns the emission of light from an excited electronic state of an atom or molecular species. A luminescence phenomenon that occurs when a chemical reaction triggers the excitation of an electronic state in a molecular species, that decays emitting light, is named chemiluminescence (CL) [36]; luminescence caused by electrogenerated chemical excitation is named electrochemiluminescence (ECL) [37]. Another luminescence phenomenon is photoluminescence (PL), where a molecule absorbs light, and then decays to a lower energy excited electronic state emitting light with a wavelength different than that of the absorbed light. Depending on the average lifetime of the excited state, the luminescence band can either be fluorescence or phosphorescence [36].

A number of biosensors exploit these phenomena and are here reported. A CL aptasensor for Hg^{2+} detection, with a limit of detection (LOD) of 16 pM, was designed by Qi et al. [38]. The sensor is based on positively-charged gold nanoparticles (AuNPs) effect, that show catalytic properties for CL reaction of luminol and H_2O_2, and on aptamer conformation change induced by Hg^{2+}. In the absence of Hg^{2+}, the aptamer causes a weak CL signal because it wraps on positive AuNPs reducing their catalytic properties. Whereas the presence of Hg^{2+} leads to a $T–Hg^{2+}–T$ complex formation preventing the interaction between aptamer and positive AuNPs, allowing the catalytic reaction to occur (Figure 3).

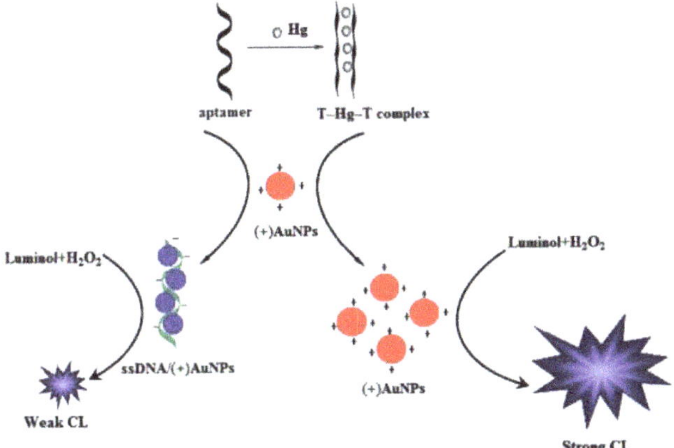

Figure 3. Metal ion-induced T–T complex mechanism for CL-based HM detection. Hg^{2+} induces $T–Hg^{2+}–T$ complex formation preventing the interaction between aptamer and positive AuNPs, allowing the catalytic reaction occurrence and a stronger CL signal emission. From ref. [38], with the permission of the Publisher.

A different Hg^{2+} biosensor, based on two label-free DNA probes and the molecular light switch complex $[Ru(phen)_2(dppz)]^{2+}$, was developed by X. Zhang et al. [39]. If Hg^{2+} is present, the two label-free DNA probes, with eight T–T mismatches, form stable DNA duplexes which allow the intercalation of $[Ru(phen)_2(dppz)]^{2+}$, leading to a significant Hg^{2+}-dependent enhancement of the luminescence intensity. A LOD of 3.5×10^{-10} M was reached.

A portable multianalyte device, based on a different recognition strategy was designed by R. K. Mishra et al. for Hg^{2+}, Pb^{2+}, and Cd^{2+} [40], obtaining a LOD of 1, 0.7, and 0.02 µg/L respectively. The device exploits a luminol–H_2O_2 mixture as a chemiluminescent system and horseradish peroxidase (HRP). The enzymatic inhibition results in a CL suppression that is analyte concentration dependent. Though, in a previous work, Deshpande et al. [41], exploiting a two enzyme based (i.e., alcohol oxidase (AlOx) and HRP) inhibition assay for single HM ion determination, showed a lower LOD (1 pg/mL) for Hg^{2+} ions.

Recently, semiconductor sensors have received significant consideration. Electrochemically-etched nano-porous silicon (PS) is considered as a promising material for luminescent chemical sensors [42,43]. Interestingly, PS layers were exploited to develop novel enzyme-based biosensor systems for determination of glucose and urea (direct) as well as HM ions (inhibitory) [44]. In particular, changes in the quantum yield of PS photoluminescence at variations in medium pH. In particular, changes in the quantum yield of PS photoluminescence at variation in medium pH is proposed for the biosensor system. The authors show that the presence of Cu^{2+}, Pb^{2+}, or Cd^{2+} ions causes an inhibition of the enzymatic reactions, resulting in a restoration of the PL quantum yield of PS. The LOD of the biosensor was approximately 10 nM. In order to develop handheld enzymatic luminescent biosensors for HMs detection, the integration of luciferase-based microfluidic chip with a portable luminometer has been also realized [45]. The LOD reached for Cu^{2+} sulfate was 2.5 mg/L.

2.1.1. Fluorescence

Transducing the molecular recognition events with the fluorescence signals is very attractive and is one of the most widely adopted methods [46]. Simultaneous measurements of multi-elements were arranged by an array-based biosensor exploiting enzymatic activity [47]. Acetylcholinesterase and urease were exploited as model enzymes and combined with a sensing probe (FITC–dextran), for the assessment of pH, urea, acetylcholine, and HMs. A LOD lowered to 10 nM was achieved for Hg^{2+} and a LOD of 50 μM was reported for Cd^{2+}.

A different kind of fluorescent transducer successfully constructed for determination of Cu^{2+} in surface water, exploits the combination of semiconductor quantum dots (QDs) and enzymatic inhibition [48]. AlOx catalyzes methanol oxidation to produce H_2O_2, inducing the quenching of QDs fluorescence. Copper ions inhibit the enzyme action and, consequently, the quenching of QDs fluorescence decreases (Figure 4b). This hybrid sensor showed a LOD of 2.75 nM.

Useful as new fluorescent sensors, carbon-based QDs (CQDs, namely biodots) have attracted growing interests thanks to their biocompatibility, chemical inertness, and water solubility. In this direction, an application of DNA-derived CQDs in metal ion sensing was demonstrated [49]. Hg^{2+} and Ag^+ are predisposed to be captured by the DNA biodots due to the existence of T and C groups (leading to T–Hg^{2+}–T or C–Ag^+–C complex), resulting in a quenched fluorescence, with the largest efficiency obtained at pH 7 and a LOD of 48 nM for Hg^{2+} and 0.31 μM· for Ag^+.

A turn-on aptasensor for Hg^{2+} detection based on graphene oxide (GO) and DNA aptamers was proposed, where GO plays a role as nano quencher (Q) to reduce the fluorescence of acridine orange (AO). The recognition process results in the simultaneous formation of T–Hg^{2+}–T and G4 structures; the formed G4 can capture AO from the GO surface, leading to fluorescence retrieval. A LOD of 0.17 nM was achieved [50].

Similarly, based on the T–Hg^{2+}–T coordination between two neighboring poly–T strands, two ready-to-use chip-based sensors match well with microarray technology for Hg^{2+} detection in the turn-on and turn-off modality [51]. The induced dislocation of the complementary poly-adenine (poly–A) strand, labeled with either a fluorophore (F) or a (Q), allows the turn-off and turn-on detection of Hg^{2+}, respectively (Figure 4c). A lower LOD was achieved in the turn-off mode (3.6 vs. 8.6 nM).

Remarkably, with the aim to remove the HM-fluorescence quenching effect, a magnetic separation was integrated for Hg^{2+} sensing based on the formation of the T–Hg^{2+}–T structure [52], allowing a LOD value of 0.2 nM.

Another multi-analyte biosensor based on parallel analysis of microarray technology was developed exploiting DNAzymes [53]. In particular, copper and lead ion-dependent DNAzymes are first associated with their corresponding DNA substrates on the surface of aldehyde-modified slides. Then, in the presence of the specific ions, the DNA cleavage of the substrate takes place, inducing a strong variation in fluorescence signal. The sensor showed a LOD value of 0.6 ppb for Cu^{2+} and 2 ppb for Pb^{2+}. A higher sensitivity for Pb^{2+}, with a LOD of 1 nM, was achieved by a similar approach, exploiting a Cy5-labeled DNA/RNA chimera (Figure 4f) as substrate [54].

Working on complex real samples, enzymatic degradation represents a threat to the structural integrity of D-DNAzymes. In this context, L-DNAzymes show similar recognition capability and catalytic capacity with respect to their enantiomer. A promising biosensor for Pb^{2+} ion detection was realized by building a Pb^{2+}-specific L-DNAzyme, allowing to obtain a LOD of 3 nM [55]. DNAzymes have also been exploited for Ag^+ detection [56]. As known, the most studied interaction between DNA and Ag^+ is the specific binding with C residues [33,57,58]. This interaction was used to develop Ag^+ biosensors [56,59,60] and for the assembling of fluorescent Ag nanoclusters [61,62]. Saran et al. [63] described the first Ag^+-specific RNA-cleaving DNAzyme, successfully integrated in the specific biosensor. A catalytic beacon biosensor is obtained by labeling the 3′ end of the DNAzyme strand with a black hole, which, upon hybridization, quenches the signal of the fluorophore located on the 5′ end of the substrate. The Ag^+-induced substrate cleavage enables fluorescence retrieval. A LOD of 24.9 nM was shown.

Even though DNAzyme-based lead sensors generally demonstrate good sensitivity, the high synthesis cost of these molecule limited their extensive application. A DNA sensor based on Pb^{2+}-stabilized G4 formation was proposed with a LOD of 3.79 ppb [64]. In the absence of Pb^{2+}, a fluorescent tracer intercalates with the single-stranded coil and strongly emits. While, in the presence of Pb^{2+}, the random-coil folds into a G4 structure leading to signal reduction (Figure 4a).

Commonly, a DNA-based biosensor for Pb^{2+} detection is frequently inclined to interference from Hg^{2+}, due to the $T-Hg^{2+}-T$ interaction between Hg^{2+} and T residues. A label-free system with a LOD in the nanomolar range was optimized (also in the presence of Hg^{2+}) based on the Pb^{2+}-induced G4 formation with cationic polythiophene water-soluble conjugated polymer (PMNT), as described in the colorimetric transduction method section of this review [65].

In another arrangement, Y.F. Zhu et al. proposed a singly-labeled bifunctional probe consisting of a Cd^{2+}-specific aptamer (CAP), capable to act as the recognition element for Cd^{2+} and the signal reporter [66]. The Cd^{2+} presence induces the switching of the CAP coil conformation to a stem-loop structure, which brings the four guanosine bases at the 5′ end close to 6-Fam at the 3′ end, resulting in fluorescence quenching. The biosensor showed a LOD of ~2 nM.

Interestingly, G4 structures have been also exploited to develop a duplex functional fluorescent biosensor for distinct detection of Pb^{2+} and Hg^{2+} [67]. A K^+-induced fluorescent G4 probe was assembled by a G-rich strand and a porphyrin. The sequence presents many T residues in addition to G residues, allowing to bind Pb^{2+} or Hg^{2+} selectively, changing into a more stably non-fluorescent G4 and a hairpin-like structure, respectively, resulting in PL reduction. LODs of 5.0 nM for Pb^{2+} and 18.6 nM for Hg^{2+} was reported.

As favorable as fluorescent nanomaterials, DNA-scaffolded silver nanoclusters (DNA–AgNCs) were successfully applied to a novel turn-on fluorescent biosensor [68]. When Pb^{2+} is present, the aptamer forms a G4 structure and the two darkish DNA/AgNCs positioned at the 3′ and 5′ terminus come closer, thus the fluorescence intensity increases [69]. A LOD as low as 3.0 nM was reported.

Light-up biosensors based on the target-induced release of fluorescence-labeled aptamer, from a complex with a Q-labeled short complementary sequence, were developed for Cd^{2+} and Pb^{2+} [70,71], with a LOD of 40 and 60.7 nM, respectively.

A label-free aptasensor approach for Cd (II) detection was independently exploited by Y. Luan et al. [72] and B. Zhou et al. [73], combining an aptamer with unmodified dsDNA-specific dye. Based on the principle that hybridization of two aptamers boosts the fluorescence engendered during the reaction, B. Zhou et al. showed that, in the absence of Cd^{2+}, SYBR green-I binds to the small groove of dsDNA (aptamer-complementary strand) establishing the dsDNA–dye complex and generating high fluorescence signal. The specific recognition and binding of aptamers with Cd^{2+} induce the release of the complementary strand from dsDNA and the aptamer conformational switching to a stem-loop structure, causing fluorescence decay (Figure 4d). A LOD of 0.34 ng/mL was reached.

Likewise, Y. Luan et al. reported that, induced by Cd^{2+} ions, the aptamer configuration changes from a random coil structure to an aptamer–Cd^{2+} complex. After the introduction of complementary strands and Pico Green dye (PG), a hybrid with the residual free aptamers that did not bind with Cd^{2+} is formed. This results in a higher PL signal (Figure 4e), allowing a higher sensitivity (LOD of 0.038 ng/mL) [72].

A comparable strategy was proposed for Pb^{2+} detection [74]. This biosensor is based on the principle that Pb^{2+} induces a structural change of G-rich thrombin aptamer from random coil to G4. This prevents its binding to the complementary sequences to form dsDNA and causes a fluorescence intensity decrease. The results showed a LOD of 1 ng/mL.

A label-free fluorescence sensing system was also developed for As^{3+} detection by the exonuclease III (Exo III)-assisted cascade target recycling amplification process [75], exhibiting a LOD of 5 ng/L. As signal indicator and sensing element, the 2-amino-5,6,7 trimethyl-1,8-naphthyridine and the triple-helix molecular switch were used, respectively. This sensor could detect other HM ions with newly-designed triple-helix molecular switch by using aptamer sequences.

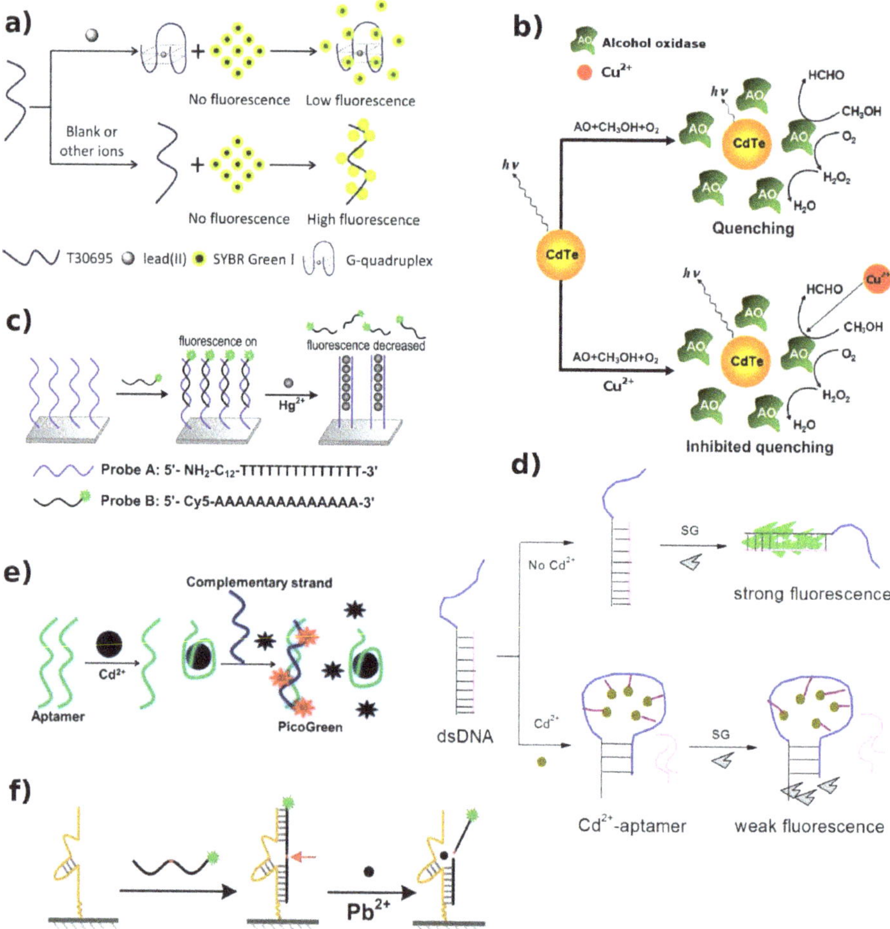

Figure 4. Various biosensing element constructs for fluorescence-based HM detection. (**a**) Pb^{2+}-stabilized G4 formation for turn off detection. In the presence of Pb^{2+}, the T30695 oligonucleotide folds into a G4 structure, leading to a PL signal reduction [64]. (**b**) Cu^{2+}-determined enzymatic inhibition for turn on detection. AO catalyzes the oxidation of methanol to hydrogen peroxide, inducing the quenching of QDs fluorescence. Cu^{2+} ions inhibits the enzymatic activity decreasing the quenching of QDs fluorescence [48]. (**c**) Metallophilic attraction of the Hg atom in the T–Hg^{2+}–T base pair mismatch. The Hg^{2+}-induced dislocation of the complementary labeled poly–A strand allows the turn-off detection mechanism. [51]. (**d**) Cd^{2+}-induced hairpin formation. The release of the complementary strand from dsDNA and the sequence conformational switching to a stem-loop structure lead to a fluorescence decay of the signal reporter [73]. (**e**) Random coil structure to aptamer–Cd^{2+} complex. After the addition of complementary strands and PG, the residual free aptamer that did not bind with Cd^{2+} forms a hybrid with complementary strands and PG dye which results in a big fluorescent enhancement [72]. (**f**) Pb^{2+}-induced hydrolytic cleavage signal-off. The catalytic strand carries out catalytic reactions for hydrolytic scission of the substrate sequence at the rA site (red arrow). Once the substrate is broken into two pieces, it dissociates from the catalytic strand with a decrease of the surface PL intensities [54]. Adapted with the permission of the Publishers.

In the frame of functional device miniaturization, combining a microfluidic sample pre-treatment module (cation exchange resins) with a DNA aptamer immobilized photoluminescent graphene oxide QD (GOQD), a novel Pb^{2+} detection platform sensor was proposed [76], exhibiting a LOD of 0.64 nM. The DNA aptamer on the GOQD specifically captures the target (forming a G4 complex) which can trigger electron transfer from GOQD to Pb^{2+} upon UV irradiation, leading the GOQD PL quenching.

2.1.2. Electrochemiluminescence

ECL is the process through which those intermediates generated at the electrodes undergo high-energy electron transfer reactions to produce an excited state that emits light, after relaxation to a lower level [77]; the process is initiated and modulated by switching an electrode voltage [78]. ECL allows small analyte detection at sub-picomolar concentration and wide dynamic range [79].

Various strategies were recently developed, such as biosensors that rely on the formation of the $T–Hg^{2+}–T$ and $Ru(phen)_3^{2+}$ or Ru-dppz, which permitted a LOD of 20 or 5.1 pM to be achieved, respectively [80,81].

In their study, X. Zhou et al. [82] reported that Bst DNA polymerase exhibits specific behaviors on the $T–Hg^{2+}–T$ biomimetic structure. The sensor exploits the MBs-labeled primer, planned to match the region of the circular padlock probe but with two T–T mismatches at the 3′ terminus. If Hg^{2+} is introduced, the DNA polymerase reaction with rolling circular amplification (RCA) mechanism is induced. Then, the resulting RCA products hybridize with the tris (bipyridine) ruthenium (TBR)-marked probes and sensed by ECL, once they are attracted to the magnet under the electrode. A LOD of 100 pM was shown.

One more method was designed by Meng Li et al., exploiting a Pb^{2+}-specific DNAzyme, achieving a LOD of 9.6×10^{-13} M [83]. In this sensor, CdS QDs and DNAzyme with Ag/ZnO coupled structures were immobilized on agold nanodendrites-modified ITO electrode. Pb^{2+}-activated DNAzyme moves the Ag/ZnO coupled structures near the surface to catalyze the reduction of part of the H_2O_2, inducing a signal intensity reduction.

Rather than utilizing DNAzyme, L. Lu et al. [84] proposed a sensor to detect Pb^{2+} using a graphene/AuNPs-modified electrode and ssDNA labeled with CdSe QDs. When Pb^{2+} is present, the G-rich ssDNA adopts the G4 conformation, leading to a shortening of the distance between the CdSe QDs and the graphene–AuNPs nanocomposite (Figure 5). This decreases the ECL intensity, allowing for the detection with a limit of 10^{-10} mol/L.

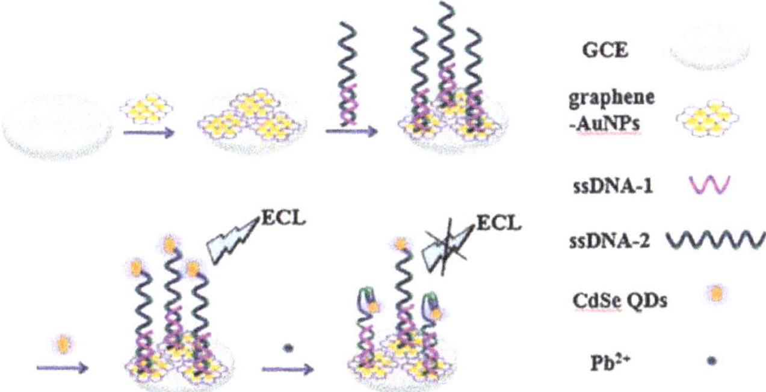

Figure 5. Metal ion-induced quadruplex construct for ECL-based HM detection The Pb^{2+} causes the G4 structure formation of the G-rich ssDNA, leading to a shortening of the distance between the CdSe QDs and the graphene–AuNPs nanocomposite, thus inducing a reduction of the ECL signal. From [84], with the permission of the Publisher.

A novel ECL sensor to detect Pb^{2+} exploiting hemin/G4-based DNAzyme on the core-shell CdSe@CdS QDs, was proposed by X.-L. Du et al. [85]. Pb^{2+}-induced G4 combines with hemin to form DNAzyme, which can catalyze H_2O_2 and oxidize 4-chloro-1-naphthol (4-CN) to form an insoluble precipitate. In the presence of Pb^{2+}, more DNAzymes are produced and, thus, more 4-CN molecules are oxidized catalytically, leading to an output signal reduction. A LOD of 0.98 fM was achieved.

Furthermore, a microfluidic paper-based device was successfully applied for concurrent detection of Pb^{2+} and Hg^{2+} based on the formation of G4 and $T–Hg^{2+}–T$ complexes, respectively [86]. Due to the different operational potentials of the two exploited labels (Si@CNCs and Ru@AuNPs), Pb^{2+} and Hg^{2+} can be quantified with a LOD of 10 pM and 0.2 nM, correspondingly.

2.2. Colorimetric Method

In colorimetric sensors, the analyte detection occurs by means of a color change of the sensing element. Current technology based on colorimetry focuses on cost reduction, miniaturization, and in-situ detection. Generally, the recognition mechanism is based on molecular interaction on the substrate surface modified with NPs and functional groups [87].

For instance, DNA adsorption by citrate-capped AuNPs could be a function of DNA conformation. DNAs without stable secondary structures allow higher colloidal stability of AuNPs against salt-induced aggregation, because they are more efficiently adsorbed. A sensor exploiting Tl^+-induced DNA folding and AuNPs was described by Hoang et al. [88]. The presence of Tl^+ inhibits the DNA adsorption by AuNPs due to G4 sequence folding. Then, adding NaCl solution, a red-to-blue color change is observable because of NPs aggregation. A LOD of 4.6 μM was achieved.

Similarly, a specific Pb^{2+}-induced G4 oligonucleotide (TBAA) probe and the cationic polythiophene (PMNT) readily form an electrostatic PMNT–TBAA red colored complex [65]. This sensor can detect Pb^{2+} traces at the micromolar level with the naked eye. Moreover, the authors report that, in the presence of Hg^{2+}, the TBAA sequence (having adenine base) has a higher selectivity for Pb^{2+} than TBA (without adenine base in the sequence). As already reported, the same biosensor exhibits a lower LOD, when working in fluorometric mode.

In order to detect Hg^{2+}, Zhu et al. [89] designed a sensor established on ssDNA, phthalic diglycol diacrylate (PDDA) and AuNPs. The $T–Hg^{2+}–T$ structure is much stronger than the interchain contact between ssDNA and PDDA. When the ssDNA recognizes Hg^{2+}, a random coil-to-hairpin structure change occurs, avoiding ssDNA interaction with PDDA. Therefore, the free PDDA induces AuNP aggregation (Figure 6b), displaying a color change as a function of Hg^{2+} concentration. The LOD was as low as 5 nM.

A multianalyte responsive sensor, able to identify Ag^+, Hg^{2+}, Cr^{3+}, Sn^{4+}, Cd^{2+}, Pb^{2+}, Zn^{2+}, and Mn^{2+} was designed by Tan et al. [90]. It is based on differential colorimetric and fluorescent response of FAM-DNA-AuNP once conjugated to a specific metal ion. A LOD of 50 nM was achieved.

A different approach for visual detection of Hg^{2+}, Ag^+, Cu^{2+}, Cd^{2+}, Pb^{2+}, Cr^{6+}, and Ni^{2+} was reported by Hossain and Brennan [91]. An enzymatic reaction is optimized on a sol gel matrix-spotted bioactive paper device; β-galactosidase-substrate catalysis produces a colorimetric signal intensity, which is inversely proportional to the metal ion amount. The sensitivity was different for the diverse ions, as reported in Table 1.

Another paper device was designed by J. Xu et al. [92], for the detection of Pb^{2+} via colorimetric and ECL techniques, exploiting a metal-specific DNAzyme and rGO–PdAu–GOx labeled oligonucleotide hybrid. The dual mode sensor showed a lower LOD in the ECL readout (0.14 nM) than in the colorimetric one (LOD: 1.6 nM).

A sensor based on mushroom apo-tyrosinase, entrapped in polyacrylamide gel, was developed by Kaur and Verma [93] in order to detect Cu^{2+}, which act as enzyme cofactor for levodopa (L-DOPA) conversion, with a corresponding color change. The shown LOD was 0.01 ug/L.

A colorimetric Hg^{2+} detection was also optimized on a test strip, exploiting biotin-labeled and thiolated DNA-modified AuNPs and a T-rich DNA immobilized on the nitrocellulose membrane. Under optimized conditions, the LOD achieved for Hg^{2+} was 3 nM [94] or 5 nM [95].

Another colorimetric paper-based platform, involving $T–Hg^{2+}–T$ coordination chemistry and AuNPs aggregation, showed a LOD of 50 nM [96]. In a different way, a linker T-rich DNA and sequences complementary to the AuNPs DNA was designed to induce particle aggregation [97]. Remarkably, Hg^{2+} ions induce the linker DNA folding, allowing AuNPs to quickly disassemble and return to red color. A lower LOD (5.4 nM) was shown with respect to the AuNPs aggregation strategy described.

In another work concerning disposable lateral flow strips, the authors examined hairpin probe-modified AuNPs and $T–Hg^{2+}–T$ structure-based strategy. An additional T-rich, digoxin-labeled DNA strand was considered in order to hybridize with $T–Hg^{2+}–T$ coordination. Then, digoxin dsDNA–AuNPs complexes are captured by immunoreaction with the anti-digoxin Ab immobilized (Figure 6a) on the strip and revealed by a red band [98]. Interestingly, a lower LOD for Hg^{2+} was shown (0.1 nM), when compared with non-immunochromatographic approaches.

On parallel route, an Exo III-catalyzed target recycling approach was employed to improve the sensitivity of a similar disposable strip on the basis of Hg^{2+}-triggered toehold binding. Using AuNPs as the tracer enables the detection of Hg^{2+} with a LOD of ~1 pM [99]. Moreover, in order to sense Cu^{2+}, a lateral flow device based on specific ion-dependent DNA-cleaving DNAzyme and AuNPs was developed, achieving a LOD of 10 nM [100].

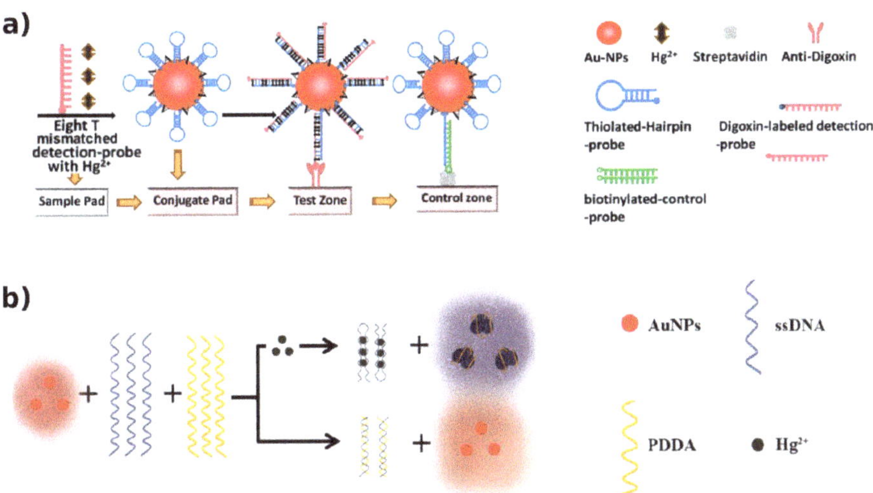

Figure 6. $T–Hg^{2+}–T$ structure and hairpin probe-modified AuNPs-based strategy for colorimetry-based HM detection. (**a**) The digoxin dsDNA–AuNPs complexes are captured by immunoreaction with the anti-digoxin Ab* immobilized on the strip and revealed by a red band [98]. (**b**) With the formation of the $T–Hg^{2+}–T$, a random coil-to-hairpin structure change occurs, avoiding ssDNA interaction with PDDA. A color change is observed due to the AuNP aggregation by free PDDA [89]. Adapted with the permission of the Publishers.

2.3. Evanescent Wave

This method employs the evanescent field of an optical fiber to excite the biological recognition molecule, producing a fluorescence signal. An optical fiber is essentially a cylindrical dielectric waveguide with an inner core having a refractive index greater than that of the cladding. EW exploits the phenomenon of total internal reflection (i.e., the transition of light in the optical fiber by continually reflecting off the cladding–core interface without data loss. If the cladding is removed, the evanescent field can interact with the fiber surroundings [101]. In order to immobilize biological recognition elements on the optical fiber surface, various methods have been reported [102], such as direct or mediated covalent immobilization, adsorption, or entrapment in polymer matrices.

Based on this transduction, a DNAzyme-based sensor for Pb^{2+} detection was developed by N. Yildirim et al. [103], using GR-5 Pb^{2+}-dependent DNAzyme. In the presence on Pb^{2+}, the active molecule can catalyze the cleavage of an RNA base embedded in the fluorescent-labeled (Cy5.5) DNA substrate. After that, the released Cy5.5-labeled fragments hybridize with the complementary strands immobilized on the optical fiber, and Pb^{2+} detection is revealed by PL signal. Restored over 100 times, this sensor showed no important performance decay and a LOD of 1.03 nM.

Another sensor based on DNAzyme for Pb^{2+} detection was realized by R. Wang et al. [104] The whole sensing procedure requires three steps: (i) Pb^{2+} ion determines the cleavage of the DNA substrate at the single RNA site, by the DNAzyme, causing the release of a short ssDNA arm which will be used in the second step; (ii) the released ssDNA hybridizes with a complementary DNA strand immobilized on MBs in solution, causing the competitive detachment of the originally hybridized probes (streptavidin–ssDNA–Cy5.5) and the remaining dsDNA–MBs complex is removed by magnetic separation; (iii) the released signal probe is pumped into the flow cell of the biosensing platform, where it can be captured by the desthiobiotin-modified fiber. A LOD of 1 nM was achieved, with the possibility to be reused at least 250 times.

To avoid the use of MBs and to keep the same performance, a similar system was optimized for Hg^{2+} sensing [105], by introducing a quencher. Fluorescein-labeled DNA strands with streptavidin (DNA–SA) were designed to hybridize with Q-labeled cDNA strands (Q–DNA). Hg^{2+} induces an enhancement in the PL signal because of the Q-DNA release, once the DNA-SA is bound to form a hairpin structure stabilized by the T–T mismatch (Figure 7a). A LOD of 1.06 nM was shown.

With a similar LOD, F. Long et al. [106] developed another sensor based on T–T mismatch pairs and fluorescently (Cy5.5)-labeled cDNA. The DNA probe has two functional elements: a T–T mismatch pair that can bind with Hg^{2+} ions, and a short sequence that can hybridize with the fluorescently-labeled cDNA. Via a structure competitive mode, Hg^{2+} ions lead to a decrease of the signal. The authors stated a LOD of 2.1 nM, with a reproducibility over 100 times.

In more recent works, the same author proposed two structure-switching DNA optical biosensors for detection of HM ions [107,108]. The developed approaches for Hg^{2+} or Pb^{2+} detection, respectively, differ by the FNA-based strategy exploited (i.e., T–T mismatch or G4 aptamer (Figure 7b)).

Once introduced in the modified optofluidic cell, the specific metal ion-induced aptamer conformation change reduces the binding of a fluorescently-labeled free DNA with the immobilized DNA probe, causing a decrease of fluorescence signal. A LODs of 1.2 nM for Hg^{2+} and 0.22 nM for Pb^{2+} were reported. One more sensor for Hg^{2+} and Pb^{2+} detection, based on T–T mismatch-containing DNA or DNAzyme, respectively, was developed by S. Han et al. [109]. In this system, the detection of HM contaminants is carried out exploiting two complementary DNA sequences, one labeled with a Cy3.3 and one with a Q. The metal ion induces structural modification, causing paired-strand dehybridization and, consequently, the binding of the Cy3.3-labeled segment to the cDNA probe on the fiber surface. By excitation via EW, a detectable fluorescence signal is generated, with a LOD of 22 pM for Hg^{2+} and 20 nM for Pb^{2+}.

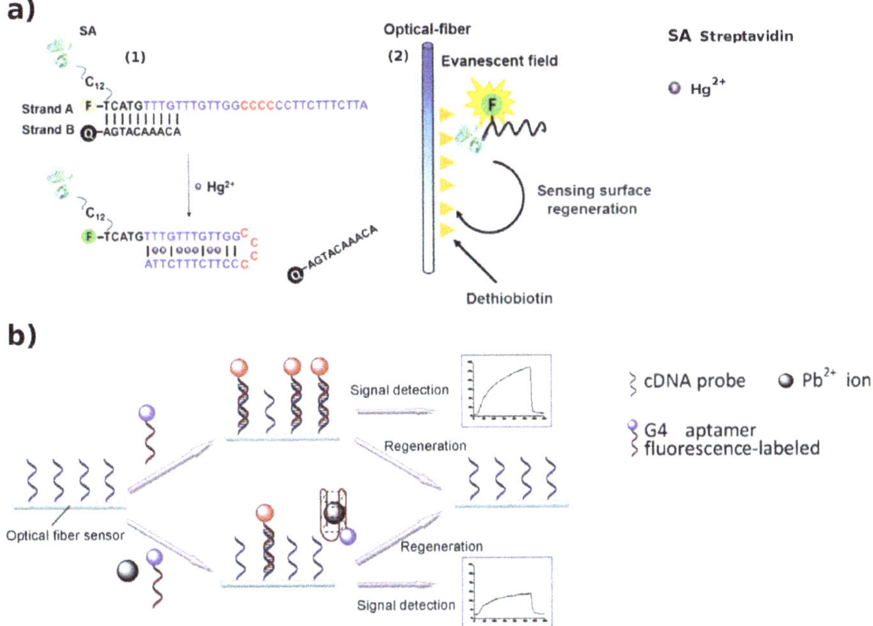

Figure 7. FNAs constructs for evanescent wave-based HM detection (**a**) (1) T–T mismatch-driven biosensing by triple functional DNA–protein conjugates for facile detection of mercury ions; (2) Once the DNA–SA is bound to form a hairpin structure stabilized by the T–T mismatch, an enhancement in the signal is observed [105]. (**b**) G4-driven lead ions biosensing. A decrease of fluorescence signal is recorded by the Pb^{2+}-induced aptamer conformation change (G4) that reduces the binding of the fluorescently-labeled free DNA with the immobilized complementary strand [108]. Adapted with the permission of the Publishers.

2.4. Surface-Enhanced Raman Spectroscopy

In SERS, definite metallic surfaces are used to intensify Raman scattering of the specific element, by benefitting from localized surface plasmon resonances. Noteworthy: (i) SERS spectra can provide information about the chemical structure of the target, (ii) it permits rapid detection; (iii) weak Raman scattering of water makes its background signal negligible [110].

A highly sensitive DNAzyme-centered SERS quadratic amplification method, based on a bio-barcode and hybridization chain reaction, was developed for Pb^{2+} detection [111]. The system includes a DNAzyme-MB complex, a SERS active bio-barcode (composed of the capture probe matching the stem of hairpin DNA, and Raman dye-labeled DNA) on top of AuNPs, to produce a strong SERS signal. Adding Pb^{2+}, once a DNA–Pb^{2+} complex is formed, a catalytic cleavage of the substrate sequence takes place, giving rise to a series of reaction steps, finally leading to quantitative Pb^{2+} detection with a LOD of 70 fM. The method can be further applied to different elements by substituting the lead-responsive DNAzyme with the specific functional DNA.

Combining a specific As^{3+} aptamer, a reporter molecule and Raman-labeled Au@Ag core–shell NPs, a novel SERS strategy was proposed [112]. In the absence of As^{3+} ions, the aptamer and the reporter are absorbed on Au@Ag; while when they are present, the As^{3+} ions compete with NPs for binding to the aptamer, inducing its release from the NP surface, which then aggregate generating SERS "hot spots". This amplification strategy allowed to obtain a LOD of 0.1 ppb.

Likewise, a label-free SERS device was developed for sensing of Hg^{2+} [113], exploiting aptamer-derivatized SiO_2@Au core–shell NPs. The DNA aptamer consists of two segments, one containing guanine (G) and adenine (A) bases as signal reporters and the other segment, with consecutive T, as the Hg^{2+} recognition element. The single strand poly–T shows a flexible structure; when present, Hg^{2+} ions cause the formation of T–Hg^{2+}–T pairs via N–Hg^{2+}–N J-coupling bonding. As a result, the DNA molecule adopts vertical alignment (Figure 8a), changing respective Raman intensities of A and G bases in the sequence. In this system, Hg^{2+} detection showed a limit of 10 nM.

The system formerly suggested by L. Zhang et al., requiring a fluorescent label, resulted as more sensitive [114]. It was based on nanoporous gold as the plasmonic surface and a Cy5-labeled aptamer as the optical tag for Hg^{2+} detection. The coordination of a pair of poly–T oligos with the metal ion induces the molecule perpendicular arrangement, as above described. Consequently, an amplification of the fluorophore surface-enhanced resonance Raman scattering signal (SERRS) variation is observed. A LOD of 1 pM was reported.

Exploiting the same Hg^{2+} biorecognition element, W. Ma et al. obtained SERS-enhanced Hg^{2+} detection, thanks to the T–Hg^{2+}–T-induced assembly of gold nanostar dimers [115]. A great number of "hotspots" were formed, inducing an increase of electromagnetic field over an extensive connecting region. A LOD of 0.8 pg/mL was reached, showing a higher sensitivity if compared with the similar strategy exploited for As^{3+} detection (Figure 8b) described in [112].

A selective single nanowire-on-film (SNOF) sensor for Hg^{2+} was realized exploiting the hybridization between T-rich ssDNAs and complementary Cy5-labeled DNAs [116]. In the presence of Hg^{2+}, T-rich DNAs fold into hairpin structures to form T–Hg^{2+}–T pairs, leading to an easy release of Cy5-tagged DNAs. The free-labeled ssDNAs are then caught by the SNOF derivatized with cDNAs, turning on the SERRS signal. A LOD of 100 pM was achieved.

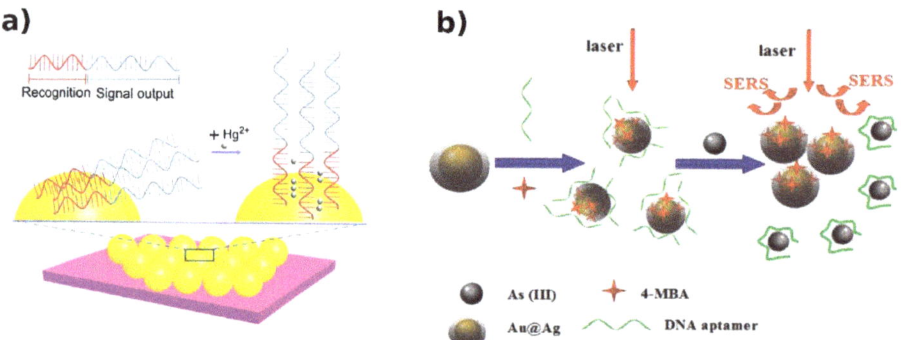

Figure 8. FNAs constructs for SERS-based HM detection. (**a**) Hg^{2+} causes a vertical alignment of DNA molecules due to the formation of T–Hg^{2+}–T pairs via N–Hg^{2+}–N J-coupling bonding, changing respective Raman intensities [113]. (**b**) As^{3+} induces the aptamer release from NP surface, inducing NP aggregation and the generation of SERS "hot spots" [112]. Adapted with permission of the Publishers.

2.5. Förster Resonance Energy Transfer

FRET is a physical process where a non-radiative energy transfer from an excited state molecule (donor) to another molecule (acceptor) occurs, by means of intermolecular long-range dipole–dipole coupling. When the energy transfer takes place from donor to acceptor, the fluorescence intensity of the donor decreases. An essential requirement for effective transfer is that an overlap exists between the fluorescence spectrum of the donor and the absorbance spectrum of the acceptor. The rate and the efficiency of the energy transfer depends on the sixth power of the distance between donor and acceptor [117]. Various combinations of donor–acceptor pairs have been used, such as two fluorophores, fluorophore with AuNP, fluorophore with an intercalator or with a dark absorber [118].

For instance, T–Hg^{2+}–T complex-induced conformational change of ssDNA allows one-step sensing of Hg^{2+} in a AuNPs-based sensor developed by G. Wang et al. [119]. The AuNPs were used as acceptor and FAM as donor. The DNA probes tagged with a FAM on 3′ and thiol on 5′ end were bound to AuNPs. In order to enable an enhanced FRET process, FAM and AuNPs need to be close to each other, as occurs when the conformation of the DNA probe changes into a hairpin structure leading to fluorescence signal quenching. A LOD of 8 nM was achieved with this approach.

Using the catalyzed hairpin assembly technique, a different aptasensor for Hg^{2+} was developed by K. Chu-mong et al. [120]. The suggested strategy exploits a Hg^{2+} aptamer–catalyst complex and two hairpin DNA: H1—fluorescein (donor) and H2—tetramethylrhodamine (acceptor). The formation of the T–Hg^{2+}–T complex releases the catalyst strand, triggering the signal amplification step: Hairpin assembly is catalyzed turning H1 and H2 into a duplex. Consequently, FRET efficiency increases and the Hg^{2+} concentration can be measured with nanomolar LOD.

An opposite functional scheme was described for Ag^+ sensing by Y.-J. Chen et al. [121]. Fusing the cyan fluorescent protein (donor) and the yellow fluorescent protein through a truncated CupR protein. CupR contains a dimerization helix and a metal binding domain. The presence of Ag^+ ions causes the decrease in FRET efficiency by inducing conformational change of the biorecognition element (Figure 9b).

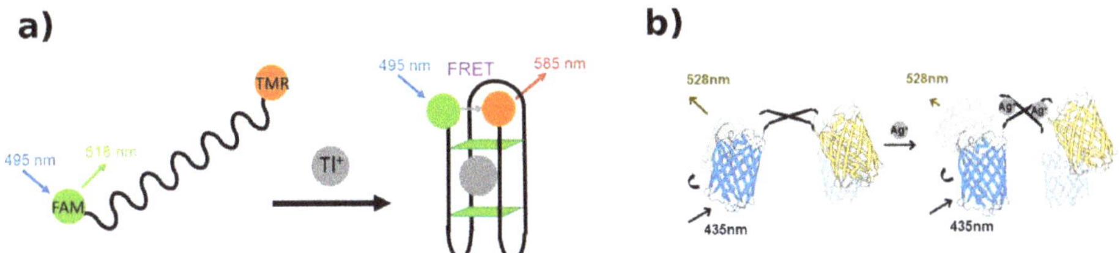

Figure 9. FNAs and protein constructs for FRET-based HM detection. (**a**) The Tl^+ causes the G4 structure formation, leading to a shortening of the distance between the donor and acceptor, thus inducing an enhancement in FRET efficiency [88]. (**b**) The decrease in FRET efficiency is induced by conformational change of the CupR protein, in the presence of Ag^+ [121]. Adapted with permission of the Publishers.

More complex systems were also designed to simultaneously detect several HMs. Using the establishment of C–Ag^+–C and T–Hg^{2+}–T complexes, Cy5 and TAMRA as acceptors and CdTe QDs as donors, C. Hao et al. [122] successfully detected Ag^+ and Hg^{2+} with a LOD of 2.5 and 1.8 nM, respectively. When a specific ion is present, if donor and acceptor are in close proximity, a fluorescence intensity increase will take place.

Interestingly, J. Xia et al. engineered specific DNA sequences for Hg^{2+}, Pb^{2+}, and Ag^+, integrating them in two DNA strands and labeling these strands with multicolor fluorophores, in order to realize a cascade FRET [123]. In this way, only one excitation wavelength is needed to obtain a fingerprint-like spectrum in multianalyte monitoring. The sensor works in a dynamic range from 100 nM to 2 μM for Ag^+ and Hg^{2+} and can detect as low as 20 nM Pb^{2+}.

As already described, M. Hoang et al. [88] demonstrated that a sensor based on G4 DNAs, FAM (donor), and TMR (acceptor) can be used for Tl^+ detection (Figure 9a) with a LOD of 59 μM, unusually lower than that showed by the colorimetric transduction method.

2.6. Surface Plasmon Resonance

When light incides on a metal surface, plasmons are generated, whose propagation is very sensitive to the variations in the material refractive index. This alteration can be caused by biomolecular interaction (probe–target) or by a structural modification of the molecules linked to the sensor surface [124,125].

For instance, the detection of Cu^{2+} was achieved by associating a SPR biosensor with the competitive adsorption of proteins [126]. The interaction between bio-receptors (native proteins (albumin)) and Cu^{2+} ions leads to protein denaturation, inducing a lower affinity between protein–gold surface, thus initiating the competitive displacement by the native one (Figure 10a), which is monitored by SPR measurement with a LOD down to 0.1 mg/L.

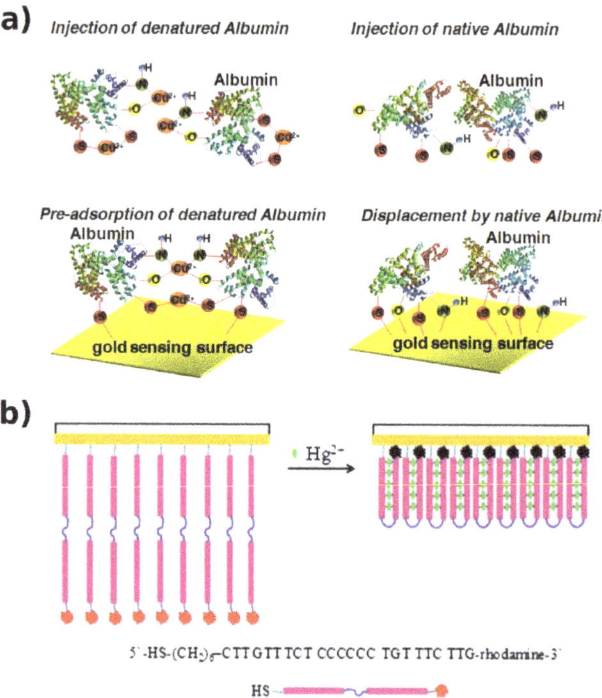

Figure 10. (**a**) Protein and DNA structure conformational changes, for SPR-based HM detection. The interaction between native proteins and Cu^{2+} leads to the protein structure denaturation and weakens its attraction on the sensing surface. The competitive displacement by the native one causes variations in the SPR angle profile [126]. (**b**) The rhodamine-labeled ssDNA folds into the T–Hg^{2+}–T-mediated hairpin loop; this structural change approaches the rhodamine fraction near to the Au surface causing the increase in the SPR signal and the PL quenching [127]. Adapted with permission of the Publishers.

A mercury (II) sensor, based on the dissociation rate of the trans-acting factor MerR from the cis-element, was investigated by SPR [128]. The sensor, modified with dsDNA including the cis-element (Pmer), can monitor the dissociation stage of MerR or protein-tagged MerR from the cis-element, enabling measurement of Hg^{2+} with a LOD of 5 µg/L.

Non-specific adsorption can influence the SPR accuracy. In this direction, a laser scanning confocal imaging and SPR were combined to realize a system for Hg^{2+} detection [127]. By adding Hg^{2+}, the rhodamine-labeled ssDNA folds into the T–Hg^{2+}–T-mediated hairpin structure and this structural change attracts the rhodamine fraction in proximity to the Au surface (Figure 10b). A double effect is observed: SPR signal heightening and PL quenching. From the PL quenching status, the strand folding is monitored in real time, and the Hg^{2+} detection is recorded by the SPR signal, as a function of refractive index and thickness variations of the Au surface, achieving a LOD of 0.01 ng/mL.

In this rich context, the summary of the recently described biosensors is schematized in Table 1. Here, biosensors are classified by their sensitivity (from lower to higher LOD) with respect to a specific ion, within the same transduction method on real samples, but with diverse bio-signaling strategies.

Moreover, in order to clearly illustrate the most sensitive recent methods as well as the bio-recognition elements giving the lowest detection limits, two comparative tables (Tables 2 and 3, respectively) are proposed and shown below. Then, a representative drawing (Figure 11) aims to show the most sensitive detection strategies, with respect to a specific analyte, applied in real samples.

Table 1. Recently reported biosensors sorted in function of the lower limit of detection (LOD), with respect to a specific analyte within the specific transduction method. Cited literature refers to analytic procedures performed on real samples.

Transduction Method	Analyte	Receptor	LOD	Linear Range	Real Sample	Reference
Luminescence	Hg^{2+}	Enzyme	1 pg/mL	(5–500) pg/mL	T.W., M.W.	[41]
Luminescence	Hg^{2+}	DMB-DNA (T–Hg^{2+}–T)	16 pM	$(6.2 \times 10^{-10} – 1.2 \times 10^{-8})$ M $(1.2 \times 10^{-8} – 1.2 \times 10^{-6})$ M	T.W., S.W.	[38]
Luminescence	Hg^{2+}	DMB-DNA (T–Hg^{2+}–T)	3.5×10^{-10} M	$(1.0 \times 10^{-9} – 1.5 \times 10^{-7})$ M	S.W.	[39]
Luminescence	Hg^{2+}	Enzyme	1 µg/L	(1–60) µg/L	S.W.	[40]
Luminescence	Pb^{2+}	Enzyme	0.7 µg/L	(0.7–54) µg/L	S.W.	[40]
Luminescence	Cd^{2+}	Enzyme	0.02 µg/L	(0.02–45) µg/L	S.W.	[40]
Fluorescence	Pb^{2+}	DMB-DNA (G4)	0.64 nM	—	T.W., S.W.	[76]
Fluorescence	Pb^{2+}	L-DNAzyme	3 nM	(5–100) nM	S.W.	[55]
Fluorescence	Pb^{2+}	DMB-DNA (G4)	3.0 nM	(5–50) nM	T.W., S.W.	[68]
Fluorescence	Pb^{2+}	DMB-DNA (G4)	1 ng m/L	1 ng m/L to over 1 mg m/L	T.W., M.W.	[74]
Fluorescence	Pb^{2+}	DMB-DNA (G4)	5.0 nM	(10–200) nM	S.W.	[67]
Fluorescence	Pb^{2+}	DMB-DNA (G4)	6 nM	(0–120) nM	T.W.	[65]
Fluorescence	Pb^{2+}	DNAzyme		—	S.W.	[53]
Fluorescence	Pb^{2+}	DMB-DNA	60.7 nM	(100–1000) nM	S.W.	[71]
Fluorescence	Hg^{2+}	DMB-DNA (T–Hg^{2+}–T; G4)	0.17 nM	(0.5–50) nM	T.W.	[50]
Fluorescence	Hg^{2+}	DMB-DNA (T–Hg^{2+}–T)	0.2 nM	(2–160) nM	S.W.	[52]
Fluorescence	Hg^{2+}	DMB-DNA (T–Hg^{2+}–T)	3.6 nM (turn-off) 8.6 nM (turn-on)	(0.01–10) µM —	M.W., milk.	[51]
Fluorescence	Hg^{2+}	Enzyme	<10 nM	—	T.W., S.W.	[47]
Fluorescence	Hg^{2+}	DMB-DNA (G4)	18.6 nM	(200–1000) nM	S.W.	[67]
Fluorescence	Hg^{2+}	DMB-DNA (T–Hg^{2+}–T)	48 nM	(0–0.5) µM (0.5–6) µM	S.W.	[49]
Fluorescence	Cd^{2+}	DMB-DNA	0.038 ng/mL	(0.10–100) µg/mL	S.W.	[72]
Fluorescence	Cd^{2+}	DMB-DNA	2.15 nM	(7.19–100) nM and 200 nM–5.0 µM	T.W., S.W., U.W.	[66]
Fluorescence	Cd^{2+}	DMB-DNA	0.34 µg/L	(1.12–224.82) µg/L	S.W., T.W., U.W.	[73]
Fluorescence	Cd^{2+}	DMB-DNA	40 nM	(0–1000) nM	S.W.	[70]
Fluorescence	Cd^{2+}	Enzyme	50 µM	—	T.W., S.W.	[47]
Fluorescence	Ag^+	RNA-cleaving DNAzyme	24.9 nM	—	S.W.	[63]
Fluorescence	Ag^+	DMB-DNA (C–Ag^+–C)	0.31 µM	(0–10) µM	S.W.	[49]
Fluorescence	As^{3+}	DMB-DNA	5 ng/L	10 ng/L–10 mg/L	T.W., S.W.	[75]
Fluorescence	Cu^{2+}	Enzyme	0.176 ng/mL	(0–2.4) ng/mL	S.W.	[48]
Fluorescence	Cu^{2+}	DNAzyme	0.6 ppb	—	S.W.	[53]
Fluorescence	Cu^{2+}	Enzyme	—	—	T.W., S.W.	[47]

Table 1. *Cont.*

Transduction Method	Analyte	Receptor	LOD	Linear Range	Real Sample	Reference
ECL	Hg^{2+}	DMB-DNA (T–Hg^{2+}–T)	5.1 pM	$(5.0 \times 10^{-11}$–$1.0 \times 10^{-8})$ M	T.W., S.W.	[81]
ECL	Hg^{2+}	DMB-DNA (T–Hg^{2+}–T) and DNA polymerase	100 pM	—	T.W., S.W.	[82]
ECL	Hg^{2+}	DMB-DNA (T–Hg^{2+}–T; G4)	0.2 nM	$(5.0 \times 10^{-10}$–$1.0 \times 10^{-6})$ M	S.W.	[86]
ECL	Pb^{2+}	G4 based DNAzyme	0.98 fM	$(1.0 \times 10^{-15}$–$1.0 \times 10^{-11})$ M	S.W.	[85]
ECL	Pb^{2+}	DNAzyme	9.6×10^{-13} M	$(5.0 \times 10^{-12}$–$4.0 \times 10^{-6})$ M	S.W.	[83]
ECL	Pb^{2+}	DMB-DNA (C–Pb^{2+}–C; G4)	10 pM	$(3 \times 10^{-11}$–$1.0 \times 10^{-6})$ M	S.W.	[86]
ECL	Pb^{2+}	DMB-DNA (G4)	10^{-10} mol/L	$(10^{-8}$–$10^{-10})$ mol/L	S.W.	[84]
ECL	Pb^{2+}	DNAzyme; Enzyme	0.14 nM	(0.5–2000) nM	T.W., S.W.	[92]
Colorimetry	Pb^{2+}	DNAzyme	1.6 nM	(5–2000) nM	T.W., S.W.	[92]
Colorimetry	Pb^{2+}	DMB-DNA (G4)	—	—	T.W.	[65]
Colorimetry	Cu^{2+}	Enzyme	0.01 µg/L	(0.1–25) µg/L	M.W., milk	[93]
Colorimetry	Hg^{2+}	DMB-DNA (T–Hg^{2+}–T); Enzyme	1 pM	—	S.W.	[99]
Colorimetry	Hg^{2+}	DMB-DNA (T–Hg^{2+}–T); Ab*	0.1 nM	(0.1–100) nM	S.W.	[98]
Colorimetry	Hg^{2+}	DMB-DNA (T–Hg^{2+}–T)	0.15 nM (UV-vis spectroscopy) 5 nM (naked eye)	(0.25–500) nM (UV-vis spectroscopy)	S.W.	[89]
Colorimetry	Hg^{2+}	DMB-DNA (T–Hg^{2+}–T)	3 nM	—	S.W.	[94]
Colorimetry	Hg^{2+} Ag^{+} Cu^{2+} Cd^{2+} Pb^{2+} Cr^{6+} Ni^{2+}	Enzyme	0.001 ppm 0.002 ppm 0.020 ppm 0.020 ppm 0.140 ppm 0.150 ppm 0.230 ppm	—	T.W., S.W.	[91]
Colorimetry	Hg^{2+}	DMB-DNA (T–Hg^{2+}–T)	5.4 nM	(0–1500) nM	S.W.	[97]
Colorimetry	Hg^{2+}	DMB-DNA (T–Hg^{2+}–T)	50 nM	—	S.W.	[96]
Colorimetry	Tl^{+}	DMB-DNA (G4)	4.6 µM	—	S.W.	[88]
Colorimetry	Ag^{+} Hg^{2+} Cr^{3+} Sn^{4+} Cd^{2+} Pb^{2+} Zn^{2+} Mn^{2+}	DMB-DNA	~50 nM	—	S.W.	[90]

Table 1. *Cont.*

Transduction Method	Analyte	Receptor	LOD	Linear Range	Real Sample	Reference
EW	Pb^{2+}	DMB-DNA	0.22 nM	(1.0–300) nM	M.W., T.W.	[108]
EW	Pb^{2+}	DNAzyme; Protein	1 nM	(20–800) nM	T.W., M.W.	[104]
EW	Pb^{2+}	DNAzyme	1.03 nM	—	T.W.	[103]
EW	Pb^{2+}	DNAzyme	20 nM	—	E.W.	[109]
EW	Hg^{2+}	DMB-DNA (T–Hg^{2+}–T)	22 pM	22 pM–10 nM	E.W.	[109]
EW	Hg^{2+}	DMB-DNA (T–Hg^{2+}–T); Protein	1.06 nM	(75–1000) nM	S.W., M.W., T.W.	[105]
EW	Hg^{2+}	DMB-DNA (T–Hg^{2+}–T); cDNA	2.1 nM	—	M.W., T.W.	[106]
SERS	As^{3+}	DMB-DNA	0.1 ppb	(0.5–10) ppb	S.W.	[112]
SERS	Pb^{2+}	DNAzyme	70 fM	0.1 pM–0.1 μM	T.W., R.W.	[111]
SERRS	Hg^{2+}	DMB-DNA	1 pM	—	U.W.	[114]
SERS	Hg^{2+}	DMB-DNA (T–Hg^{2+}–T)	0.8 pg/mL	(0.002–1) ng m/L	T.W.	[115]
SERS	Hg^{2+}	DMB-DNA (T–Hg^{2+}–T)	10 nM	$(1 \times 10^{-8}$–$1 \times 10^{-3})$ M	U.W., S.W.	[113]
FRET	Hg^{2+}	DMB-DNA (T–Hg^{2+}–T)	1.8 nM	—	T.W., L.W.	[122]
FRET	Hg^{2+}	DMB-DNA (T–Hg^{2+}–T)	(7.03 ± 0.18) nM	(10.0–200.0) nM	S.W.	[120]
FRET	Hg^{2+}	DMB-DNA (T–Hg^{2+}–T)	8 nM	(20–90) nM	T.W.	[119]
FRET	Ag^{+}	DMB-DNA (C–Ag^{+}–C)	2.5 nM	—	T.W., S.W.	[122]
FRET	Ag^{+}	Protein	—	—	T.W., S.W.	[121]
FRET	Tl^{+}	DMB-DNA (G4)	59 μM	—	S.W.	[88]
LSCI-SPR	Hg^{2+}	DMB-DNA (T–Hg^{2+}–T)	0.01 ng/mL	(0.01–100) ng/mL	T.W.	[127]
SPR	Hg^{2+}	DMB-DNA	5 μg/L	$(10^{1}$–$10^{4})$ μg/L	M.W.	[128]
SPR	Cu^{2+}	Protein	~0.1 mg/L	—	T.W., M.W.	[126]

Tap water (T.W.); mineral water (M.W.); surface water (S.W.); underground water (U.W.); environmental water (E.W.). Direct metal binding DNA sequence (DMB-DNA); Antibody (Ab*) (*indirectly exploited).

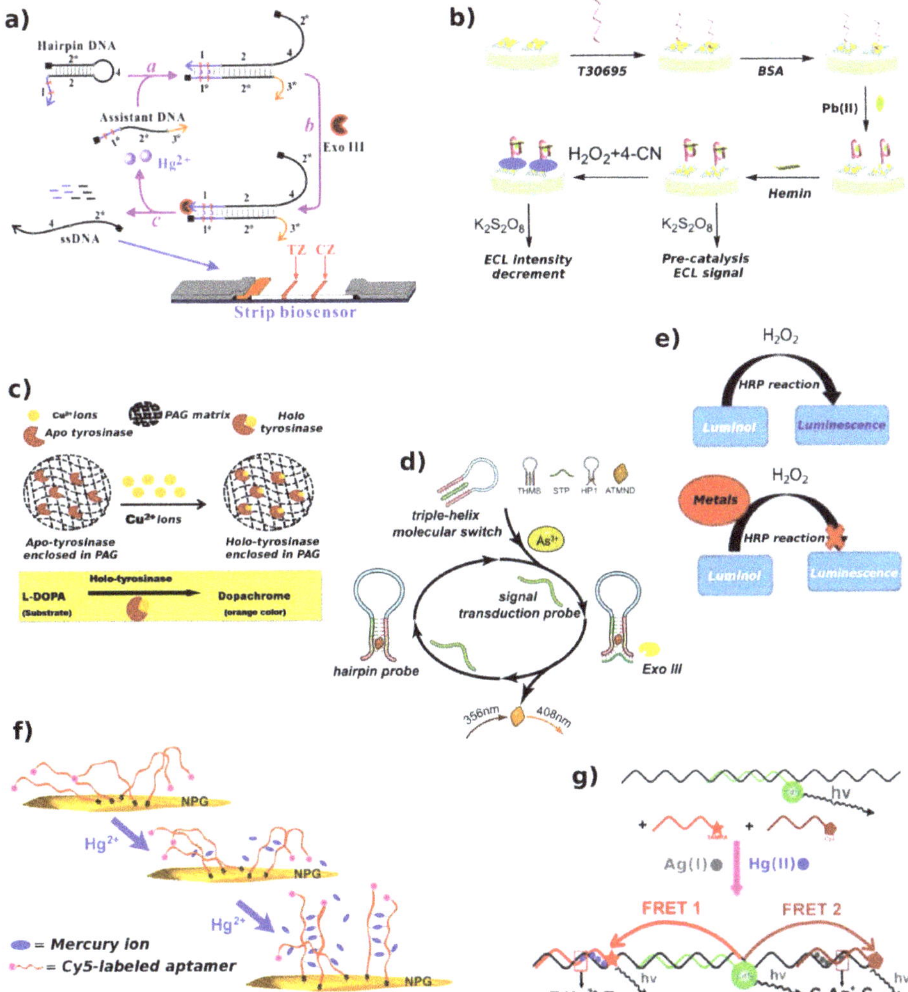

Figure 11. Most sensitive biorecognition strategy/specific HM ion, in real samples. (**a**) Disposable strip biosensor based on Hg²⁺-induced toehold binding and Exo III-assisted signal amplification [99]; (**b**) ECL Pb²⁺ sensor based on hemin/G4-based DNAzyme biocatalysis [85]; (**c**) Cu²⁺ triggered conversion of apo-tyrosinase disc into holo-tyrosinase one, and consequent L-DOPA to dopachrome transformation [93]; (**d**) As³⁺ detection by Exo III-assisted cascade target-recycling amplification scheme [75]; (**e**) Possible on-site analysis of HMs by means of the HRP-based bioassay [40]; (**f**) Aptamer-modified NPG-based SERRS sensing of Hg²⁺ [114]; (**g**) Ag+ and Hg²⁺ detection by FRET between QD and organic dyes [122]. Adapted with permission of the Publishers.

Table 2. Most sensitive transduction methods, with respect to the specific analyte (n. of published works $\geq$ 2), applied in real samples.

Analyte	Transduction Method	LOD	Reference
Pb^{2+}	ECL	0.98 fM	[85]
Hg^{2+}	SERRS	1 pM	[114]
Hg^{2+}	Colorimetry	1 pM	[99]
Cd^{2+}	Luminescence	0.02 µg/L	[40]
As^{3+}	Fluorescence	5 ng/L	[75]
Cu^{2+}	Colorimetry	0.01 µg/L	[93]
Ag^{+}	FRET	2.5 nM	[122]

Table 3. Bio-recognition elements giving the lowest detection limits, within the same transduction method, with respect to a specific analyte (n. of published works $\geq$ 3).

Transduction Method	Analyte	Signaling Strategy	LOD	Reference
Luminescence	Hg^{2+}	Enzyme	1 pg/mL	[41]
Fluorescence	Pb^{2+}	G4 aptamer–GOQD	0.64 nM	[76]
Fluorescence	Hg^{2+}	AO–DNA Aptamer (T–Hg^{2+}–T; G4)	0.17 nM	[50]
Fluorescence	Cd^{2+}	Aptamer/cDNA	0.038 ng/mL	[72]
Fluorescence	Cu^{2+}	Alcohol Oxidase inhibition	0.176 ng/mL	[48]
ECL	Pb^{2+}	Hemin/G4-based DNAzyme	0.98 fM	[85]
ECL	Hg^{2+}	Hg^{2+}-specific AuNP–ssDNA/cDNA	5.1 pM	[81]
Colorimetric	Hg^{2+}	Hairpin DNA(T–Hg^{2+}–T)/Exonuclease III	1 pM	[99]
Colorimetric	Pb^{2+}	DNAzyme/GO–PdAu–(GOx)–ssDNA	1.6 nM	[92]
EW	Pb^{2+}	Cy5.5–G4 aptamer	0.22 nM	[108]
EW	Hg^{2+}	Quencher T-rich DNA/ Cy3–cDNA/ssDNA probe (T–Hg^{2+}–T)	22 pM	[109]
SERRS	Hg^{2+}	Cy5–Aptamer–NPG (T–Hg^{2+}–T)	1 pM	[114]
FRET	Hg^{2+}	TAMRA–ssDNA/QD–ssDNA (T–Hg^{2+}–T)	1.8 nM	[122]

3. Conclusions

Nucleic acids, biocatalysts, antibodies, receptors, etc., are natural or biomimetic elements with distinctive features such that they have been engaged as recognition probes since the first public biosensor description in a paper, over 55 years ago, in which Dr. L. C. Clark termed his device as an "enzyme electrode" [129]. In the fields of environmental and food analysis, water and milk exemplify the matrices involved in potential HM ion contamination. In this context, although most of the developed systems were tested only on buffered solutions, plenty of optical biosensors appropriate for real samples showed up in the last decade for possible environmental and food quality monitoring applications. Continuous advances are presented, exploiting nano-microtechnology and biotechnology, such as for miniaturization of integrated systems, genetic engineering of receptors, enzymes, and microorganisms, as well upgrading of bioelement immobilization methods.

Thus far, a number of metals can be selectively sensed by DNA sequences down to the low ppb level [11]. Accordingly, Table 3 shows that direct metal binding DNA sequences allow obtainment of the highest sensitivity. In detail, the biorecognition mechanisms more frequently adopted are those based on T–T mismatch and G-quadruplex, respectively, for Hg^{2+} and Pb^{2+}; nonetheless, to the same extent, functional nucleic acids (DNAzyme) are exploited for Pb^{2+}. Among the optical biosensors here reviewed, those applied in real samples, namely milk and water (specifically tap water, mineral water, surface water, underground water), have been assessed by spike test, largely for Hg^{2+}, Pb^{2+}, and Cd^{2+} ions, in descending order, and, in small part, also for Cu^{2+}, Ag^{+}, Cr^{3+}, As^{3+}, Tl^{+}, and Sn^{4+} ions, as summarized in Table 1. Remarkably, ten multianalyte optical devices (able to sense up to eight HM ions) were shown out of a total of more than seventy biosensors here considered, with a large part of them designed to quantitatively discriminate between two ions.

Author Contributions: A.A. and R.R. collaborated to design the manuscript. A.A., A.D.T., A.D.B. and R.R. wrote the manuscript. A.A., A.D.T. and R.R. revised the manuscript.

Funding: This research was funded by [Regione Puglia, Italy] INNOLABS HYDRO RISK LAB Project (grant number NRNABW5).

Acknowledgments: A.D.B., A.A. and R.R. are grateful to MIUR for a PhD grant financed by PON project "Dottorati innovativi con caratterizzazione industriale" (Code: project n.1 DOT1312457).

Conflicts of Interest: The authors declare no conflicts of interest.

References

1. Odobašić, A.; Šestan, I.; Sabina Begić, S. Biosensors for Determination of Heavy Metals in Waters. In *Environmental Biosensors*; IntechOpen: London, UK, 2019.

2. Thévenot, D.R.; Buck, R.P.; Cammann, K.; Durst, R.A.; Toth, K.; Wilson, G.S. Electrochemical biosensors: Recommended definitions and classification (Technical Report). *Pure Appl. Chem.* **1999**, *71*, 2333–2348. [CrossRef]

3. Kang, H.; Lin, L.; Rong, M.; Chen, X. A cross-reactive sensor array for the fluorescence qualitative analysis of heavy metal ions. *Talanta* **2014**, *129*, 296–302. [CrossRef] [PubMed]

4. Wu, P.; Zhao, T.; Wang, S.; Hou, X. Semiconductor quantum dots-based metal ion probes. *Nanoscale* **2014**, *6*, 43–64. [CrossRef] [PubMed]

5. Duffus, J.H. "Heavy metals"—A meaningless term? (IUPAC Technical Report). *Pure Appl. Chem.* **2002**, *74*, 793–807. [CrossRef]

6. Kanellis, V.G. Sensitivity limits of biosensors used for the detection of metals in drinking water. *Biophys. Rev.* **2018**, *10*, 1415–1426. [CrossRef] [PubMed]

7. Wang, L.; Ma, W.; Xu, L.; Chen, W.; Zhu, Y.; Xu, C.; Kotov, N.A. Nanoparticle-based environmental sensors. *Mater. Sci. Eng. R* **2010**, *70*, 265–274. [CrossRef]

8. Aragay, G.; Josefina Pons, J.; Merkoc, A. Recent Trends in Macro-, Micro-, and Nanomaterial-Based Tools and Strategies for Heavy-Metal Detection. *Chem. Rev.* **2011**, *111*, 3433–3458. [CrossRef] [PubMed]

9. Malitesta, C.; Di Masi, S.; Mazzotta, E. From electrochemical biosensors to biomimetic sensors based on molecularly imprinted polymers in environmental determination of heavy metals. *Front. Chem.* **2017**, *5*, 47. [CrossRef]

10. Catia Algieri, C.; Drioli, E.; Guzzo, L.; Donato, L. Bio-Mimetic Sensors Based on Molecularly Imprinted Membranes. *Sensors* **2014**, *14*, 13863–13912. [CrossRef]

11. Zhou, W.; Saran, R.; Liu, J. Metal Sensing by DNA. *Chem. Rev.* **2017**, *117*, 8272–8325. [CrossRef]

12. Olaoye, O.O.; Manderville, R.A. Aptamer Utility in Sensor Platforms for the Detection of Toxins and Heavy Metals. *J. Toxins* **2017**, *4*, 12–23.

13. Zhang, J.; Sun, X.; Wu, J. Heavy Metal Ion Detection Platforms Based on a Glutathione Probe: A Mini Review. *Appl. Sci.* **2019**, *9*, 489. [CrossRef]

14. Upadhyay, L.S.B.; Nishant Verma, N. Enzyme Inhibition Based Biosensors: A Review. *Anal. Lett.* **2013**, *46*, 225–241. [CrossRef]

15. Gutiérrez, J.C.; Amaro, F.; Martín-González, A. Heavy metal whole-cell biosensors using eukaryotic microorganisms: An updated critical review. *Front. Microbiol.* **2015**, *6*, 48. [PubMed]

16. Martins, T.D.; Ribeiro, A.C.C.; de Camargo, H.S.; da Costa Filho, P.A.; Cavalcante, H.P.M.; Dias, L.D. New Insights on Optical Biosensors: Techniques, Construction and Application. In *State of the Art in Biosensors General Aspects*; Rinken, T., Ed.; IntechOpen: London, UK, 2013.

17. Cobbett, C.; Goldsbrough, P. Phytochelatins and metallothioneins: Roles in heavy metal detoxification and homeostasis. *Annu. Rev. Plant Biol.* **2002**, *53*, 159–182. [CrossRef] [PubMed]

18. Gordon, W.I.; Swee, N.T.; Stillman, M.J. A Simple Metallothionein-Based Biosensor for Enhanced Detection of Arsenic and Mercury. *Biosensors* **2017**, *7*, 14.

19. Mattiasson, B.; Danielsson, B.; Hermansson, C.; Mosbach, K. Enzyme thermistor analysis of heavy metal ions with use of Immobilized urease. *FEBS Lett.* **1978**, *85*, 203–206. [CrossRef]

20. Othman, A.; Karimi, A.; Andreescu, S. Functional nanostructures for enzyme based biosensors: Properties, fabrication and applications. *J. Mater. Chem. B* **2016**, *4*, 7178–7203. [CrossRef]

21. Reardan, D.T.; Meares, C.F.; Goodwin, D.A.; McTigue, M.; David, G.S.; Stone, M.R.; Leung, J.P.; Bartholomew, R.M.; Frincke, J.M. Antibodies against metal chelates. *Nature* **1985**, *316*, 265–268. [CrossRef]
22. Delehanty, J.B.; Jones, R.M.; Bishop, T.C.; Blake, D.A. Identification of important residues in metal-chelate recognition by monoclonal antibodies. *Biochemistry* **2003**, *42*, 14173–14183. [CrossRef]
23. Blake, D.A.; Blake, R.C., II; Abboud, E.R.; Li, X.; Yu, H.; Kriegel, A.M.; Khosraviani, M.; Darwish, I.A. Antibodies to Heavy Metals: Isolation, Characterization, and Incorporation into Microplate-Based Assays and Immunosensors. In *Immunoassay and Other Bioanalytical Techniques*, 1st ed.; van Emon, J.M., Ed.; CRC Press: Boca Raton, FL, USA, 2006; pp. 93–111.
24. Blake, D.A.; Jones, R.M.; Blake, R.C., II; Pavlov, A.R.; Darwish, I.A.; Yu, H. Antibody-based sensors for heavy metal ions. *Biosens. Bioelectron.* **2001**, *16*, 799–809. [CrossRef]
25. Liu, J.; Cao, Z.; Lu, Y. Functional Nucleic Acid Sensors. *Chem. Rev.* **2009**, *109*, 1948–1998. [CrossRef] [PubMed]
26. Zhan, S.; Wu, Y.; Wang, L.; Zhan, X.; Zhou, P. A mini-review on functional nucleic acids-based heavy metal ion detection. *Biosens. Bioelectron.* **2016**, *86*, 353–368. [CrossRef] [PubMed]
27. Iliuk, A.B.; Hu, L.; Tao, W.A. Aptamer in Bioanalytical Applications. *Anal. Chem.* **2011**, *83*, 4440–4452. [CrossRef] [PubMed]
28. Yüce, M.; Ullah, N.; Budak, H. Trends in aptamer selection methods and applications. *Analyst* **2015**, *140*, 5379–5399. [CrossRef] [PubMed]
29. Mascini, M.; Palchetti, I.; Tombelli, S. Nucleic Acid and Peptide Aptamers: Fundamentals and Bioanalytical Aspects. *Angew. Chem. Int. Ed.* **2012**, *51*, 1316–1332. [CrossRef] [PubMed]
30. Song, K.-M.; Lee, S.; Ban, C. Aptamers and Their Biological Applications. *Sensors* **2012**, *12*, 612–631. [CrossRef]
31. Sun, C.; Ou, X.; Cheng, Y.; Zhai, T.; Liu, B.; Lou, X.; Xia, F. Coordination-induced structural changes of DNA-based optical and electrochemical sensors for metal ions detection. *Dalton Trans.* **2019**, *48*, 5879–5891. [CrossRef]
32. Miyake, Y.; Togashi, H.; Tashiro, M.; Yamaguchi, H.; Oda, S.; Kudo, M.; Tanaka, Y.; Kondo, Y.; Sawa, R.; Fujimoto, T.; et al. Mercury^{II}-Mediated Formation of Thymine-Hg^{II}-Thymine Base Pairs in DNA Duplexes. *J. Am. Chem. Soc.* **2006**, *128*, 2172–2173. [CrossRef]
33. Ono, A.; Cao, S.; Togashi, H.; Tashiro, M.; Fujimoto, T.; Machinami, T.; Oda, S.; Miyake, Y.; Okamoto, I.; Tanaka, Y. Specific interactions between silver(I) ions and cytosine–cytosine pairs in DNA duplexes. *Chem. Commun.* **2008**, *39*, 4825–4827. [CrossRef]
34. Gellert, M.; Lipsett, M.N.; Davies, D.R. Helix formation by guanylic acid. *Proc. Natl. Acad. Sci. USA* **1962**, *48*, 2013–2018. [CrossRef]
35. Mehta, J.; Bhardwaj, S.K.; Bhardwaj, N.; Paul, A.K.; Kumar, P.; Kim, K.H.; Deep, A. Progress in the biosensing techniques for trace-level heavy metals. *Biotechnol. Adv.* **2016**, *34*, 47–60. [CrossRef]
36. Omary, M.A.; Patterson, H.H. Luminescence, Theory. In *Encyclopedia of Spectroscopy and Spectrometry*, 2nd ed.; Lindon, J.C., Tranter, G.E., Koppenaal, D.W., Eds.; Elsevier: Amsterdam, The Netherlands, 2010; pp. 1372–1391.
37. Kulmala, S.; Suomi, J. Current status of modern analytical luminescence methods. *Anal. Chim. Acta* **2003**, *500*, 21–69. [CrossRef]
38. Qi, Y.; Xiu, F.-R.; Yu, G.; Huang, L.; Li, B. Simple and rapid chemiluminescence aptasensor for Hg^{2+} in contaminated samples: A new signal amplification mechanism. *Biosens. Bioelectron.* **2017**, *87*, 439–446. [CrossRef]
39. Zhang, X.; li, Y.; Su, H.; Zhang, S. Highly sensitive and selective detection of Hg^{2+} using mismatched DNA and a molecular light switch complex in aqueous solution. *Biosens. Bioelectron.* **2010**, *25*, 1338–1343. [CrossRef]
40. Mishra, R.K.; Rhouati, A.; Bueno, D.; Anwar, M.W.; Shahid, S.A.; Sharma, V.; Marty, J.L.; Hayat, A. Design of portable luminescence bio-tool for on-site analysis of heavy metals in water samples. *Int. J. Environ. Anal. Chem.* **2018**, *98*, 1081–1094. [CrossRef]
41. Deshpande, K.; Mishra, R.K.; Bhand, S. A High Sensitivity Micro Format Chemiluminescence Enzyme Inhibition Assay for Determination of Hg(II). *Sensors* **2010**, *10*, 6377–6394. [CrossRef]
42. Sailor, M.J. *Porous Silicon in Practice: Preparation, Characterization and Applications*; Wiley-VCH: Weinheim, Germany, 2012; pp. 1–38.
43. Salcedo, W.J.; Fernandez, F.J.R.; Rubim, J.C. Photoluminescence quenching effect on porous silicon films for gas sensors application. *Spectrochim. Acta* **2004**, *60*, 1065–1070. [CrossRef]

44. Syshchyk, O.; Skryshevsky, V.A.; Soldatkin, O.O.; Soldatkin, A.P. Enzyme biosensor systems based on porous silicon photoluminescence for detection of glucose, urea and heavy metals. *Biosens. Bioelectron.* **2015**, *66*, 89–94. [CrossRef]
45. Lukyanenko, K.A.; Denisov, I.A.; Sorokin, V.V.; Yakimov, A.S.; Esimbekova, E.N.; Belobrov, P.I. Handheld Enzymatic Luminescent Biosensor for Rapid Detection of Heavy Metals in Water Samples. *Chemosensors* **2019**, *7*, 16. [CrossRef]
46. Strianese, M.; Staiano, M.; Ruggiero, G.; Labella, T.; Pellecchia, C.; D'Auria, S. Fluorescence-Based Biosensors. In *Spectroscopic Methods of Analysis. Methods in Molecular Biology (Methods and Protocols)*; Bujalowski, W., Ed.; Humana Press: Totowa, NJ, USA, 2012; Volume 875, pp. 193–216.
47. Tsai, H.C.; Doong, R.A. Simultaneous determination of pH, urea, acetylcholine and heavy metals using array-based enzymatic optical biosensor. *Biosens. Bioelectron.* **2005**, *20*, 1796–1804. [CrossRef]
48. Guo, C.; Wang, J.; Cheng, J.; Zhifei, Z. Determination of trace copper ions with ultrahigh sensitivity and selectivity utilizing CdTe quantum dots coupled with enzyme inhibition. *Biosens. Bioelectron.* **2012**, *36*, 69–74. [CrossRef]
49. Song, T.; Zhu, X.; Zhou, S.; Yang, G.; Gan, W.; Yuan, Q. DNA derived fluorescent bio-dots for sensitive detection of mercury and silver ions in aqueous solution. *Appl. Surf. Sci.* **2015**, *347*, 505–513. [CrossRef]
50. Guo, H.; Li, J.; Li, Y.; Wu, D.; Ma, H.; Wei, Q.; Du, B.; Guo, H.; Li, J.; Li, Y.; et al. A turn-on fluorescent sensor for Hg^{2+} detection based on graphene oxide and DNA aptamers. *New J. Chem.* **2018**, *42*, 11147–11152. [CrossRef]
51. Du, J.; Liu, M.; Lou, X.; Zhao, T.; Wang, Y.; Xue, Z.; Zhao, J.; Xu, Y. Highly Sensitive and Selective Chip-Based Fluorescent Sensor for Mercuric Ion: Development and Comparison of Turn-On and Turn-Off Systems. *Anal. Chem.* **2012**, *84*, 8060–8066. [CrossRef]
52. Sun, C.; Sun, R.; Chen, Y.; Tong, Y.; Zhu, J.; Bai, H.; Zhang, S.; Zheng, H.; Ye, H. Utilization of aptamer-functionalized magnetic beads for highly accurate fluorescent detection of mercury (II) in environment and food. *Sens. Actuators B* **2018**, *255*, 775–780. [CrossRef]
53. Zuo, P.; Yin, B.C.; Ye, B.C. DNAzyme-based microarray for highly sensitive determination of metal ions. *Biosens. Bioelectron.* **2009**, *25*, 935–939. [CrossRef]
54. Liu, M.; Lou, X.; Du, J.; Guan, M.; Wang, J.; Ding, X.; Zhao, J. DNAzyme-based fluorescent microarray for highly selective and sensitive detection of lead(II). *Analyst* **2012**, *137*, 70–72. [CrossRef]
55. Liang, H.; Xie, S.; Cui, L.; Wub, C.; Zhang, X. Designing a biostable L-DNAzyme for lead(II) ion detection in practical samples. *Anal. Methods* **2016**, *8*, 7260–7264. [CrossRef]
56. Li, T.; Shi, L.; Wang, E.; Dong, S. Multifunctional G-Quadruplex Aptamers and Their Application to Protein Detection. *Chem. Eur. J.* **2009**, *15*, 3347–3350. [CrossRef]
57. Zheng, Y.; Yang, C.; Yang, F.; Yang, X. Real-Time Study of Interactions between Cytosine–Cytosine Pairs in DNA Oligonucleotides and Silver Ions Using Dual Polarization Interferometry. *Anal. Chem.* **2014**, *86*, 3849–3855. [CrossRef]
58. Urata, H.; Yamaguchi, E.; Nakamura, Y.; Wada, S.-I. Pyrimidine–pyrimidine base pairs stabilized by silver(I) ions. *Chem. Commun.* **2011**, *47*, 941–943. [CrossRef]
59. Freeman, R.; Finder, T.; Willner, I. Multiplexed analysis of Hg^{2+} and Ag^{+} ions by nucleic acid functionalized CdSe/ZnS quantum dots and their use for logic gate operations. *Angew. Chem. Int. Ed.* **2009**, *48*, 7818–7821. [CrossRef]
60. Wen, Y.Q.; Xing, F.F.; He, S.J.; Song, S.P.; Wang, L.H.; Long, Y.T.; Li, D.; Fan, C.H. A graphene-based fluorescent nanoprobe for silver(I) ions detection by using graphene oxide and a silver-specific oligonucleotide. *Chem. Commun.* **2010**, *46*, 2596–2598. [CrossRef]
61. Su, Y.-T.; Lan, G.-Y.; Chen, W.-Y.; Chang, H.-T. Detection of Copper Ions Through Recovery of the Fluorescence of DNA-Templated Copper/Silver Nanoclusters in the Presence of Mercaptopropionic Acid. *Anal. Chem.* **2010**, *82*, 8566–8572. [CrossRef]
62. Richards, C.I.; Choi, S.; Hsiang, J.-C.; Antoku, Y.; Vosch, T.; Bongiorno, A.; Tzeng, Y. L.; Dickson, R.M. Oligonucleotide-Stabilized Ag Nanocluster Fluorophores. *J. Am. Chem. Soc.* **2008**, *130*, 5038–5039. [CrossRef]
63. Saran, R.; Liu, J. A Silver DNAzyme. *Anal. Chem.* **2016**, *88*, 4014–4020. [CrossRef]
64. Zhan, S.; Wu, Y.; Luo, Y.; Liu, L.; He, L.; Xing, H.; Zhou, P. Label-free fluorescent sensor for lead ion detection based on lead(II)-stabilized G-quadruplex formation. *Anal. Biochem.* **2014**, *462*, 19–25. [CrossRef]

65. Lu, Y.; Li, X.; Wang, G.; Wen Tang, W. A highly sensitive and selective optical sensor for Pb^{2+} by using conjugated polymers and label-free oligonucleotides. *Biosens. Bioelectron.* **2013**, *39*, 231–235. [CrossRef]

66. Zhu, Y.-F.; Wang, Y.-S.; Zhou, B.; Yu, J.-H.; Peng, L.-L.; Huang, Y.-Q.; Li, X.-J.; Chen, S.-H.; Tang, X.; Wang, X.-F. A multifunctional fluorescent aptamer probe for highly sensitive and selective detection of cadmium (II). *Anal. Bioanal. Chem.* **2017**, *409*, 4951–4958. [CrossRef]

67. Zhu, Q.; Liua, L.; Xinga, Y.; Zhou, X. Duplex functional G-quadruplex/NMM fluorescent probe for label-free detection of lead(II) and mercury(II) ions. *J. Hazard. Mater.* **2018**, *355*, 50–55. [CrossRef]

68. Zhang, B.; Wei, C. Highly sensitive and selective detection of Pb^{2+} using a turn-on fluorescent aptamer DNA silver nanoclusters sensor. *Talanta* **2018**, *182*, 125–130. [CrossRef]

69. Yin, B.; Ma, J.; Le, H.-N.; Wang, S.; Xu, Z.; Ye, B. A new mode to light up an adjacent DNA-scaffolded silver probe pair and its application for specific DNA detection. *Chem. Commun.* **2014**, *50*, 15991–15994. [CrossRef]

70. Wang, H.; Cheng, H.; Wang, J.; Xu, L.; Chen, H.; Pei, R. Selection and characterization of DNA aptamers for the development of light-up biosensor to detect Cd(II). *Talanta* **2016**, *154*, 498–503. [CrossRef]

71. Chen, Y.; Li, H.; Gao, T.; Zhang, T.; Xu, L.; Wang, B.; Wang, J.; Pei, R. Selection of DNA aptamers for the development of light-up biosensor to detect Pb(II). *Sens. Actuators B* **2018**, *254*, 214–222. [CrossRef]

72. Luan, Y.; Lu, A.; Chen, J.; Fu, H.; Xu, L. A label-Free Aptamer-Based Fluorescent Assay for Cadmium Detection. *Appl. Sci.* **2016**, *6*, 432. [CrossRef]

73. Zhou, B.; Yang, X.-Y.; Wang, Y.-S.; Yi, J.-C.; Zeng, Z.; Zhang, H.; Chen, Y.-T.; Hua, X.-J.; Suo, Q.-L. Label-free fluorescent aptasensor of Cd^{2+} detection based on the conformational switching of aptamer probe and SYBR green I. *Microchem. J.* **2019**, *144*, 377–382. [CrossRef]

74. Huang, Y.; Yan, J.; Fang, Z.; Zhang, C.; Bai, W.; Yan, M.; Zhu, C.; Gao, C.; Chen, A. Highly sensitive and selective optical detection of Pb^{2+} using a label-free fluorescent aptasensor. *RSC Adv.* **2016**, *6*, 9030.

75. Pan, J.; Li, Q.; Zhou, D.; Chen, J. Ultrasensitive aptamer biosensor for arsenic (III) detection based on labelfree triple-helix molecular switch and fluorescence sensing platform. *Talanta* **2018**, *189*, 370–376. [CrossRef]

76. Park, M.; Ha, H.D.; Kim, Y.T.; Jung, J.H.; Kim, S.-H.; Kim, D.H.; Seo, T.S. Combination of a Sample Pretreatment Microfluidic Device with a Photoluminescent Graphene Oxide Quantum Dot Sensor for Trace Lead Detection. *Anal. Chem.* **2015**, *87*, 10969–10975. [CrossRef]

77. Miao, W. Electrogenerated Chemiluminescence and Its Biorelated Applications. *Chem. Rev.* **2008**, *108*, 2506–2553. [CrossRef]

78. Richter, M.M. Electrochemiluminescence (ECL). *Chem. Rev.* **2004**, *104*, 3003–3036. [CrossRef]

79. Hu, L.; Xu, G. Applications and trends in electrochemiluminescence. *Chem. Soc. Rev.* **2010**, *39*, 3275–3304. [CrossRef]

80. Tang, C.-X.; Zhao, Y.; He, X.-W.; Yin, X.-B. A 'turn-on' electrochemiluminescent biosensor for detecting Hg^{2+} at femtomole level based on the intercalation of $Ru(phen)_3^{2+}$ into ds-DNA. *Chem. Commun.* **2010**, *46*, 9022–9024. [CrossRef]

81. Huang, R.-F.; Liu, H.-X.; Gai, Q.-Q.; Liu, G.-J.; Wei, Z. A facile and sensitive electrochemiluminescence biosensor for Hg^{2+} analysis based on a dual-function oligonucleotide probe. *Biosens. Bioelectron.* **2015**, *71*, 194–199. [CrossRef]

82. Zhou, X.; Su, Q.; Xing, D. An electrochemiluminescent assay for high sensitive detection of mercury (II) based on isothermal rolling circular amplification. *Anal. Chim. Acta* **2012**, *713*, 45–49. [CrossRef]

83. Li, M.; Kong, Q.; Bian, Z.; Ma, C.; Ge, S.; Zhang, Y.; Yu, J.; Yan, M. Ultrasensitive detection of lead ion sensor based on gold nanoden-drites modified electrode and electrochemiluminescent quenching of quantum dots by electrocatalytic silver/zinc oxide coupled structures. *Biosens. Bioelectron.* **2015**, *65*, 176–182. [CrossRef]

84. Lu, L.; Guo, L.; Li, J.; Kang, T.; Cheng, S. Electrochemiluminescent detection of Pb^{2+} by graphene/gold nanoparticles and CdSe quantum dots. *Appl. Surf. Sci.* **2016**, *388*, 431–436. [CrossRef]

85. Du, X.-L.; Kang, T.-F.; Lu, L.-P.; Cheng, S.-Y. An electrochemiluminescence sensor based on CdSe@CdS functionalized MoS_2 and hemin/G-quadruplex-based DNAzyme biocatalytic precipitation for sensitive detection of Pb(II). *Anal. Methods* **2018**, *10*, 51–58. [CrossRef]

86. Zhang, M.; Ge, L.; Ge, S.; Yan, M.; Yu, J.; Huang, J.; Liu, S. Three-dimensional paper-based electrochemiluminescence device for simultaneous detection of Pb^{2+} and Hg^{2+} based on potential-control-technique. *Biosens. Bioelectron.* **2013**, *41*, 544–550. [CrossRef]

87. VS, A.P.; Joseph, P.; SCG, K.D.; Lakshmanan, S.; Kinoshita, T.; Muthusamy, S. Colorimetric sensor for rapid detection of various analytes. *Mater. Sci. Eng. C* **2017**, *78*, 1231–1245.

88. Hoang, M.; Huang, P.-J.J.; Liu, J. G-Quadruplex DNA for Fluorescent and Colorimetric Detection of Thallium(I). *ACS Sens.* **2016**, *1*, 137–143. [CrossRef]

89. Zhu, Y.; Cai, Y.; Zhu, Y.; Zheng, L.; Ding, J.; Quan, Y.; Wang, L.; Qi, B. Highly sensitive colorimetric sensor for Hg^{2+} detection based on cationic polymer/DNA interaction. *Biosens. Bioelectron.* **2015**, *69*, 174–178. [CrossRef]

90. Tan, L.; Chen, Z.; Zhao, Y.; Wei, X.; Li, Y.; Zhang, C.; Wei, X.; Hu, X. Dual channel sensor for detection and discrimination of heavy metal ions based on colorimetric and fluorescence response of the AuNPs-DNA conjugates. *Biosens. Bioelectron.* **2016**, *85*, 414–421. [CrossRef]

91. Hossain, S.M.Z.; Brennan, J.D. β-Galactosidase-Based Colorimetric Paper Sensor for Determination of Heavy Metals. *Anal. Chem.* **2011**, *83*, 8772–8778. [CrossRef]

92. Xu, J.; Zhang, Y.; Li, L.; Kong, Q.; Zhang, L.; Ge, S.; Yu, J. Colorimetric and Electrochemiluminescence Dual-Mode Sensing of Lead Ion Based on Integrated Lab-on-Paper Device. *ACS Appl. Mater. Interfaces* **2018**, *10*, 3431–3440. [CrossRef]

93. Kaur, G.; Verma, N. Colorimetric determination of Cu^{2+} ions in water and milk by apo-tyrosinase disc. *Sens. Actuators B* **2018**, *263*, 524–532. [CrossRef]

94. Guo, Z.; Duan, J.; Yang, F.; Li, M.; Hao, T.; Wang, S.; Wei, D. A test strip platform based on DNA-functionalized gold nanoparticles for on-site detection of mercury (II) ions. *Talanta* **2012**, *93*, 49–54. [CrossRef]

95. Duan, J.; Guo, Z.Y. Development of a test strip based on DNA-functionalized gold nanoparticles for rapid detection of mercury (II) ions. *Chin. Chem. Lett.* **2012**, *23*, 225–228. [CrossRef]

96. Chen, G.-H.; Chen, W.-Y.; Yen, Y.-C.; Wang, C.-W.; Chang, H.-T.; Chen, C.-F. Detection of Mercury(II) Ions Using Colorimetric Gold Nanoparticles on Paper-Based Analytical Devices. *Anal. Chem.* **2014**, *86*, 6843–6849. [CrossRef]

97. Torabi, S.-F.; Lu, Y. Small-molecule diagnostics based on functional DNA nanotechnology: A dipstick test for Mercury. *Faraday Discuss.* **2011**, *149*, 125–135. [CrossRef]

98. He, Y.; Zhang, X.; Zeng, K.; Zhang, S.; Baloda, M.; Gurung, A.S.; Liu, G. Visual detection of Hg^{2+} in aqueous solution using gold nanoparticles and thymine-rich hairpin DNA probes. *Biosens. Bioelectron.* **2011**, *26*, 4464–4470. [CrossRef]

99. Chen, J.; Zhou, S.; Wen, J. Disposable Strip Biosensor for Visual Detection of Hg^{2+} Based on Hg^{2+}-Triggered Toehold Binding and Exonuclease III-Assisted Signal Amplification. *Anal. Chem.* **2014**, *86*, 3108–3114. [CrossRef]

100. Fang, Z.; Huang, J.; Lie, P.; Xiao, Z.; Ouyang, C.; Wu, Q.; Wu, Y.; Liu, G.; Zeng, L. Lateral flow nucleic acid biosensor for Cu^{2+} detection in aqueous solution with high sensitivity and selectivity. *Chem. Commun.* **2010**, *46*, 9043–9045. [CrossRef]

101. Leung, A.; Shankar, P.M.; Mutharasan, R. A review of fiber-optic biosensors. *Sens. Actuators B* **2007**, *125*, 688–703. [CrossRef]

102. Taitt, C.R.; Anderson, G.P.; Ligler, F.S. Evanescent wave fluorescence biosensors. *Biosens. Bioelectron.* **2005**, *20*, 2470–2487. [CrossRef]

103. Yildirim, N.; Long, F.; He, M.; Gao, C.; Shi, H.-C.; Gu, A.Z. A portable DNAzyme-based optical biosensor for highly sensitive and selective detection of lead (II) in water sample. *Talanta* **2014**, *129*, 617–622. [CrossRef]

104. Wang, R.; Zhou, X.; Shi, H. Triple functional DNA–protein conjugates: Signal probes for Pb^{2+} using evanescent wave-induced emission. *Biosens. Bioelectron.* **2015**, *74*, 78–84. [CrossRef]

105. Wang, R.; Zhou, X.; Shi, H.; Luo, Y. T–T mismatch-driven biosensor using triple functional DNA-protein conjugates for facile detection of Hg^{2+}. *Biosens. Bioelectron.* **2016**, *78*, 418–422. [CrossRef]

106. Long, F.; Gao, C.; Shi, H.C.; He, M.; Zhu, A.N.; Klibanov, A.M.; Gu, A.Z. Reusable evanescent wave DNA biosensor for rapid, highly sensitive, and selective detection of mercury ions. *Biosens. Bioelectron.* **2011**, *26*, 4018–4023. [CrossRef]

107. Long, F.; Zhu, A.; Shi, H.; Wang, H.; Liu, J. Rapid on-site/in-situ detection of heavy metal ions in environmental water using a structure-switching DNA optical biosensor. *Sci. Rep.* **2013**, *3*, 2308. [CrossRef]

108. Long, F.; Zhu, A.; Wang, H. Optofluidics-based DNA structure-competitive aptasensor for rapid on-site detection of lead(II) in an aquatic environment. *Anal. Chim. Acta* **2014**, *849*, 43–49. [CrossRef]

109. Han, S.; Zhou, X.; Tang, Y.; He, M.; Zhang, X.; Shi, H.; Xiang, Y. Practical, highly sensitive, and regenerable evanescent-wave biosensor for detection of Hg^{2+} and Pb^{2+} in water. *Biosens. Bioelectron.* **2016**, *80*, 265–272. [CrossRef]

110. Sharma, B.; Frontiera, R.R.; Henry, A.-I.; Ringe, E.; Van Duyne, R.P. SERS: Materials, applications, and the future. *Mater. Today* **2012**, *15*, 16–25. [CrossRef]

111. Tian, A.; Liu, Y.; Gao, J. Sensitive SERS detection of lead ions via DNAzyme based quadratic signal amplification. *Talanta* **2017**, *171*, 185–189. [CrossRef]

112. Song, L.; Mao, K.; Zhou, X.; Hu, J. A novel biosensor based on Au@Ag core–shell nanoparticles for SERS detection of arsenic (III). *Talanta* **2016**, *146*, 285–290. [CrossRef]

113. Lu, Y.; Zhong, J.; Yao, G.; Huang, Q. A label-free SERS approach to quantitative and selective detection of mercury (II) based on DNA aptamer-modified SiO_2@Au core/shell nanoparticles. *Sens. Actuators B* **2018**, *258*, 365–372. [CrossRef]

114. Zhang, L.; Chang, H.; Hirata, A.; Wu, H.; Xue, Q.-K.; Chen, M. Nanoporous Gold Based Optical Sensor for Sub-ppt Detection of Mercury Ions. *ACS Nano* **2013**, *7*, 4595–4600. [CrossRef]

115. Ma, W.; Sun, M.; Xu, L.; Wang, L.; Kuang, H.; Xu, C. A SERS active gold nanostar dimer for mercury ion detection. *Chem. Commun.* **2013**, *49*, 4989–4991. [CrossRef]

116. Kang, T.; Yoo, S.M.; Yoon, I.; Lee, S.; Choo, J.; Lee, S.Y.; Kim, B. Au Nanowire-on-Film SERRS Sensor for Ultrasensitive Hg^{2+} Detection. *Chem. Eur. J.* **2011**, *17*, 2211–2214. [CrossRef]

117. Clegg, R.M. Fluorescence resonance energy transfer. *Curr. Opin. Biotechnol.* **1995**, *6*, 103–110. [CrossRef]

118. Massey, M.; Algar, W.R.; Krull, U.J. Fluorescence resonance energy transfer (FRET) for DNA biosensors: FRET pairs and Förster distances for various dye–DNA conjugates. *Anal. Chim. Acta* **2006**, *568*, 181–189. [CrossRef]

119. Wang, G.; Lu, Y.; Yan, C.; Lu, Y. DNA-functionalization gold nanoparticles based fluorescence sensor for sensitive detection of Hg^{2+} in aqueous solution. *Sens. Actuators B* **2015**, *211*, 1–6. [CrossRef]

120. Chu-mong, K.; Thammakhet, C.; Thavarungkul, P.; Kanatharana, P.; Buranachai, C. A FRET based aptasensor coupled with non-enzymatic signal amplification for mercury (II) ion detection. *Talanta* **2016**, *155*, 305–313. [CrossRef]

121. Chen, Y.-J.; Liu, C.-Y.; Tsai, D.-Y.; Yeh, Y.-C. A portable fluorescence resonance energy transfer biosensor for rapid detection of silver ions. *Sens. Actuators B* **2018**, *259*, 784–788. [CrossRef]

122. Hao, C.; Xua, L.; Xing, C.; Kuang, H.; Wang, L.; Xu, C. Oligonucleotide-based fluorogenic sensor for simultaneous detection of heavy metal ions. *Biosens. Bioelectron.* **2012**, *36*, 174–178. [CrossRef]

123. Xia, J.; Lin, M.; Zuo, X.; Su, S.; Wang, L.; Huang, W.; Fan, C.; Huang, Q. Metal Ion-Mediated Assembly of DNA Nanostructures for Cascade Fluorescence Resonance Energy Transfer-Based Fingerprint Analysis. *Anal. Chem.* **2014**, *86*, 7084–7087. [CrossRef]

124. Wijaya, E.; Lenaerts, C.; Maricot, S.; Hastanin, J.; Habraken, S.; Vilcot, J.-P. Surface plasmon resonance-based biosensors: From the development of different SPR structures to novel surface functionalization strategies. *Curr. Opin. Solid State Mater. Sci.* **2011**, *15*, 208–224. [CrossRef]

125. Asal, M.; Ozen, O.; Sahinler, M.; Polatoglu, I. Recent developments in enzyme, DNA and immuno-based biosensors. *Sensors* **2018**, *18*, 1924. [CrossRef]

126. Wang, R.; Wang, W.; Ren, H.; Chae, J. Detection of copper ions in drinking water using the competitive adsorption of proteins. *Biosens. Bioelectron.* **2014**, *57*, 179–185. [CrossRef]

127. Zhang, H.; Yang, L.; Zhou, B.; Liu, W.; Ge, J.; Wu, J.; Wang, Y.; Wang, P. Ultrasensitive and selective gold film-based detection of mercury (II) in tap water using a laser scanning confocal imaging-surface plasmon resonance system in real time. *Biosens. Bioelectron.* **2013**, *47*, 391–395. [CrossRef] [PubMed]

128. Taniguchi, M.; Siddiki, M.S.R.; Ueda, S.; Maeda, I. Mercury (II) sensor based on monitoring dissociation rate of the trans-acting factor MerR from cis-element by surface plasmon resonance. *Biosens. Bioelectron.* **2015**, *67*, 309–314. [CrossRef] [PubMed]

129. Clark, L.C.; Lyons, C. Electrode systems for continuous monitoring cardiovascular surgery. *Ann. N. Y. Acad. Sci.* **1962**, *102*, 29–45. [CrossRef] [PubMed]

Article

A Critical Comparison between Flow-Through and Lateral Flow Immunoassay Formats for Visual and Smartphone-Based Multiplex Allergen Detection

Georgina M. S. Ross [1,*], Gert IJ. Salentijn [1,2] and Michel W. F. Nielen [1,2]

[1] Wageningen Food Safety Research, Wageningen University & Research, P.O. Box 230,
   6700 AE Wageningen, the Netherlands; gert.salentijn@wur.nl (G.I.S.); michel.nielen@wur.nl (M.W.F.N.)
[2] Laboratory of Organic Chemistry, Wageningen University, Helix Building 124, Stippeneng 4,
   6708 WE Wageningen, the Netherlands
* Correspondence: georgina.ross@wur.nl

Received: 3 November 2019; Accepted: 11 December 2019; Published: 12 December 2019

Abstract: (1) Background: The lack of globally standardized allergen labeling legislation necessitates consumer-focused multiplexed testing devices. These should be easy to operate, fast, sensitive and robust. (2) Methods: Herein, we describe the development of three different formats for multiplexed food allergen detection, namely active and passive flow-through assays, and lateral flow immunoassays with different test line configurations. (3) Results: The fastest assay time was 1 min, whereas even the slowest assay was within 10 min. With the passive flow approach, the limits of detection (LOD) of 0.1 and 0.5 ppm for total hazelnut protein (THP) and total peanut protein (TPP) in spiked buffer were reached, or 1 and 5 ppm of THP and TPP spiked into matrix. In comparison, the active flow approach reached LODs of 0.05 ppm for both analytes in buffer and 0.5 and 1 ppm of THP and TPP spiked into matrix. The optimized LFIA configuration reached LODs of 0.1 and 0.5 ppm of THP and TPP spiked into buffer or 0.5 ppm for both analytes spiked into matrix. The optimized LFIA was validated by testing in 20 different blank and spiked matrices. Using device-independent color space for smartphone analysis, two different smartphone models were used for the analysis of optimized assays.

Keywords: flow-through immunoassay; lateral flow immunoassay; food allergen; multiplex; smartphone analysis; carbon nanoparticle labeling

---

1. Introduction

Food allergens are naturally occurring proteins present in a multitude of foods. Individuals with a food allergy are sensitized towards these proteins, and exposure to them can lead to adverse, sometimes life-threatening, health effects [1]. The majority of food allergen-related anaphylaxis in Europe can be attributed to peanut and tree nut allergens [2]. Allergies towards peanuts and tree nuts commonly co-exist, making the simultaneous detection of these problematic allergens desirable [3,4].

The only way for allergic individuals to avoid an allergic reaction is for them to stick to an avoidance diet. Such diets are largely reliant upon proper allergen labeling of food products. However, currently in the European Union (EU), only ingredients which have been intentionally incorporated into a food require labeling [5,6]. This means that allergens that are unintentionally present in food, such as via cross contamination, do not need to be declared, with all associated risks for allergic consumers. As a result, many food manufacturers use voluntary precautionary allergen labeling (PAL) (e.g., 'may contain' statements) in order to safeguard consumers [7].

In theory, PAL statements protect the consumer from potential allergic reactions; in reality the over-use of unregulated PAL has resulted in consumers choosing to ignore these warning statements [8].

Therefore, it is imperative to engage the public with their own food allergen analysis by developing consumer-friendly detection methods [9,10]. The cornerstones to consumer-friendly allergen detection are speed, sensitivity, ease-of-use, affordability, portability, multiplexing capability and a simple read-out system. Although some specifically consumer-oriented allergen sensors are available, such as the portable gluten and peanut sensors from NIMA, more often these biosensors are still proof-of-concept assays rather than commercial tests designed for consumers [11–14], and generally they lack multiplexing and proper validation as screening methods. A shared characteristic of novel allergen detection is the increasing trend to utilize a smartphone as an interface and readout system [9,14–17]. Using a smartphone readout improves the overall ease of result interpretation by introducing an interface that the consumer is already familiar with, alongside providing a means to wirelessly transmit results to relevant stakeholders, such as food manufacturers and restaurant personnel [18]. The Lateral Flow Immunoassay (LFIA) is widely considered the gold standard for easy-to-use, low-cost, sensitive and quick screening for food safety issues. Despite their widespread application, allergen LFIAs are often based on the analysis of a single analyte, owing to the difficulties associated with multiplexing an LFIA, including the need for careful design of test line configuration to prevent upstream detection areas from affecting downstream detection areas [19,20]. Most multiplex LFIAs for food safety focus upon the detection of low-molecular weight compounds, such as antibiotics and mycotoxins [21,22]. However, this past year has seen an increase in the development of multiplex food allergen detection LFIAs, with the development of an assay for the detection of hazelnut, ovalbumin and casein in bakery products within 10 min [23]. A further example is the multiplex, low-ppm detection of both β-lactoglobulin and β-casein, two major allergenic milk proteins, within 10 min [24].

A major drawback typically associated with LFIAs is the assay duration, which usually is 10–20 min, and is affected by mass transport limitations (MTL) and binding kinetics [25]. MTLs are caused by the fact that the target analytes need to be carried across a porous membrane, such as nitrocellulose (NC) by passive, capillary flow, and thus affect the detection speed of the assay [26]. The NC capillary flow rate is measured in the time in seconds it takes the sample front to travel 4 cm. Selection of NC based on this capillary flow rate is a compromise between assay sensitivity and assay speed with mid-speed membranes (120–150 s/4 cm) offering advantages in both areas [27]. When detection speed is not a constraint, a membrane with a slower flow rate and smaller pore size increases the available binding time between the labeled antibody–analyte and the test line antibody which can result in increased assay sensitivity [27–29]. In order to speed up LFIAs, in combination with NC with a good flow rate, antibodies with fast association rates towards their target should be used. Antibodies can be selected for their binding kinetics by in depth surface plasmon resonance (SPR)-based antibody screening and characterization. In this way a carbon nanoparticle-based hazelnut allergen LFIA has been developed, with a 30 s assay time, which as far as we know is a world record for allergen assay speed [30].

In order to overcome restrictions typically associated with LFIAs, a flow-through immunoassay format can be used instead [31,32]. Flow-through immunoassays are reported to offer the benefits of increased assay speeds, better sensitivities—owing to the use of larger sample volumes, excellent multiplexing capabilities and the absence of the 'hook-effect' [27,33,34]. The hook-effect is a phenomenon that is commonly encountered in one-step, sandwich format LFIAs. It occurs where the free analyte and the analyte which is bound to a labeled antibody compete for the limited number of binding sites available on immobilized capture antibodies, leading to a reduction in colorimetric signal and sometimes false negative results [24,35,36]. Therefore, if the correct assay working range is not determined, it could lead to consumers erroneously believing a food with a high allergen content is safe.

Flow-through assays can be prepared in different ways. Passive flow-through assays consist of LFIA materials, but in a stacked arrangement, with the membrane biofunctionalized with capture antibodies on top, and the conjugate and absorbent pads layered underneath or as flow-through enzyme-linked immunosorbent assays (ELISAs) [37–39]. An alternative flow-through approach is to insert a biofunctionalized membrane into a syringe filter holder, applying manual or mechanical pressure to the syringe to actively control the vertical flow of the reagents and the sample [40,41]. Although flow-through formats generally allow greater freedom in geometric assay design, they are prone to inter/intra-user variability [42].

The lack of agreed regulatory allergen thresholds has stalled the development of certified reference materials, preventing true comparisons to be made between various detection methods by different kit manufacturers and researchers [43]. Therefore, in this study, we use the same bioreagents to compare different geometrically designed, paper-based, flow-through and lateral flow immunoassay configurations for the simultaneous detection of hazelnut and peanut allergens with a smartphone readout system.

2. Materials and Methods

2.1. Reagents and Consumables

Washing buffer (WB) was composed of 5 mM borate buffer (BB) (pH 8.8) diluted from a mixture of 100 mM sodium tetraborate (VWR, Leuven, Belgium) and 100 mM boric acid (Merck, Darmstadt, Germany) and bovine serum albumin (BSA; Sigma-Aldrich, Zwijndrecht, the Netherlands) was added to a final concentration of 1% (*w/v*). Storage buffer (SB) consisted of 100 mM BB containing BSA to a final concentration of 1% (*w/v*). Running buffer (RB) was prepared by adding 1% BSA (*w/v*) and 0.05% Tween-20 (*v/v*) (Merck, Darmstadt, Germany) to 100 mM BB. Phosphate buffered saline (PBS; 0.01 M; pH 7.4) was purchased from Sigma-Aldrich (Sigma-Aldrich, Zwijndrecht, the Netherlands). All solutions were prepared with water from a MilliQ-system (MQ) (>18.2 MΩ/cm) purchased from Millipore (Burlington, MA, USA). 'Spezial Schwartz 4' carbon nanoparticles were purchased from Degussa AG (Frankfurt, Germany). Goat anti-mouse IgG in PBS (pH 7.6) (1.2 mg/mL; AffiniPure F(ab')$_2$ Fragment GAM IgG Fcγ) used for spraying control lines/spots was purchased from Jackson Immunoresearch Laboratories Inc. (Sanbio, Uden, the Netherlands). The hazelnut (50-6B12) and peanut (51-2A12 and 51-12D2) antibodies were developed by Wageningen Food Safety Research (WFSR), Wageningen University and Research (Wageningen, the Netherlands) according to the procedure described by Bremer et al. [44]. All antibodies were buffer exchanged from PBS (pH 7.4) into 5 mM BB (pH 8.8) using Zeba™ Spin Trap columns (Thermo Scientific; Landsmeer, the Netherlands) prior to use. Passive flow-through assays were developed from a Miriad Rapid Vertical Flow toolkit (MedMira, Halifax, NS, Canada). All active flow-through assays were developed on unbacked Whatman 0.45 µm nylon (GE Healthcare, Eindhoven, the Netherlands) 0.45 µm NC or 0.2 µm NC membranes and inserted into 13 mm Swinny syringe filter holders (Merck, Darmstadt, Germany). The assembled filter holder was attached to a 10 mL syringe (Becton-Dickinson, Utrecht, the Netherlands). Lateral flow immunoassays (LFIAs) were developed on 140 CN nitrocellulose membranes (Unisart, Sartorius, Gottinghem, Germany) secured on a plastic backing (G and L, San Jose, CA, USA) overlaid with an absorbent pad (Whatman, GE Healthcare, Eindhoven, the Netherlands). All LFIAs were heat-sealed in foil packets with silica beads and stored at room temperature until use.

2.2. Allergen Extraction

Currently, a drawback in allergen detection is that no certified, standardized reference materials are commercially available, and antigen standards and blank matrices need to be prepared in-house [45]. The influence of food processing on the protein conformation of allergens can affect their detectability [46], but this was not explicitly investigated in this study, as the focus was comparing the performance of the same antibodies applied in different immunoassay formats.

Extracts were made from hazelnuts, peanuts, blank flour, peanut-spiked flour (8 ppm) and 20 truly different biscuits (i.e., 20 different brands and varieties; see Supplementary Information, Table S1) free from peanuts/tree-nuts, which were supplied by project partners or purchased from local supermarkets. Raw hazelnuts and unsalted peanuts were frozen whole at −80 °C for 1 h. The frozen foods were homogenized using a commercial hand blender (Braun Turbo 600 W Food Processor, Braun, Oss, the Netherlands). A total protein extract was made by adding 10 mL PBS (pH 7.4) per gram of ground sample and incubating at room temperature for 1 h. Following incubation, extracts were centrifuged at $3220\times g$ for 20 min. The extracts were then filtered through a series of low protein-binding syringe filters (5 µm > 1.2 µm > 0.45 µm), and the filtrate was aliquoted and stored at −20 °C until use. To ensure sample stability, fresh aliquots were defrosted daily for experiments, and protein concentrations were determined using the NanoDrop ND 3300 (Isogen Life Sciences, De Meern, the Netherlands) prior to use. Blank biscuits were homogenized by agitating 0.5 g in a 50 mL tube with ball bearings to a fine powder. Next, 5 mL of 100 mM borate buffer was added to the tubes and agitated for 1 min with the powdered biscuit or flour. The suspension was left at room temperature for 25 min. Afterwards, extracts were filtered through a series of low protein-binding syringe filters (5 µm > 1.2 µm > 0.45 µm), aliquoted and stored at −20 °C until use. All experiments, except for matrix experiments, were performed using total hazelnut protein (THP) and total peanut protein (TPP) spiked into running buffer. For matrix experiments, 1 µL of 1000 ppm THP and TPP extract was spiked into 999 µL (*v/v*) of the 20 different blank biscuit extracts.

2.3. Carbon Black Nanoparticle Conjugation

A 1% suspension of carbon nanoparticles (CNPs) was prepared by adding 1 mL of MQ Water to 10 mg carbon and sonicating for 10 min. The resulting 1% carbon suspension was diluted five times in 5 mM BB (pH 8.8) to obtain a 0.2% suspension, which was then sonicated for 5 min. Next, 350 µL purified hazelnut or peanut antibody solution (1 mg/mL in 5 mM BB) was added to 1 mL (to make a total volume of 1.35 mL) of 0.2% carbon suspension and stirred overnight at 4 °C. The suspension was split into approximately two equal aliquots (670 µL), and 500 µL of WB was added to each before centrifuging them for 15 min at $13,636\times g$ at 4 °C. Following this, the supernatants were removed, and the pellets re-suspended in WB. This process was repeated three times. After the final wash, the supernatants were discarded, and the pellets were pooled together with 1 mL storage buffer and stored at 4 °C until use.

2.4. Multiplex Passive Flow-through

The plastic cartridge, biofunctionalized membrane and absorbent pad (absorption volume of 200 µL) from a Miriad Rapid Vertical Flow technology toolkit was used to create the passive flow-through assays. A schematic representation of the passive flow-through assay is shown in Figure 1A.

The membranes were biofunctionalized by manually depositing 0.5 µL of the peanut, hazelnut and control antibody solutions (1 mg/mL) in three distinct regions using a pipette. The tip of the pipette was touched very lightly against the membrane to dispense a consistent antibody spot. The membranes were dried for 45 min. Once dried, three drops of RB were added via a dropper bottle and allowed to saturate the membrane. Immediately after, 50 µL of the mixed allergen extract (diluted in RB; 1000 ppm, 100 ppm, 10 ppm, 1 ppm, 0.1 or 0 ppm) was pipetted dropwise onto the membrane and allowed to absorb fully. Next, a 10 µL suspension of 10 × diluted carbon labeled-monoclonal antibodies (CNP-mAbs) was pipetted onto the membrane and allowed to absorb fully. Finally, three drops of RB were applied to wash the membranes. The assays were read immediately with the naked eye and an image was acquired with a smartphone camera. LOD values for visual inspection were established at the lowest concentration that reproducibly yielded a signal that could be observed and distinguished from the background by the naked eye.

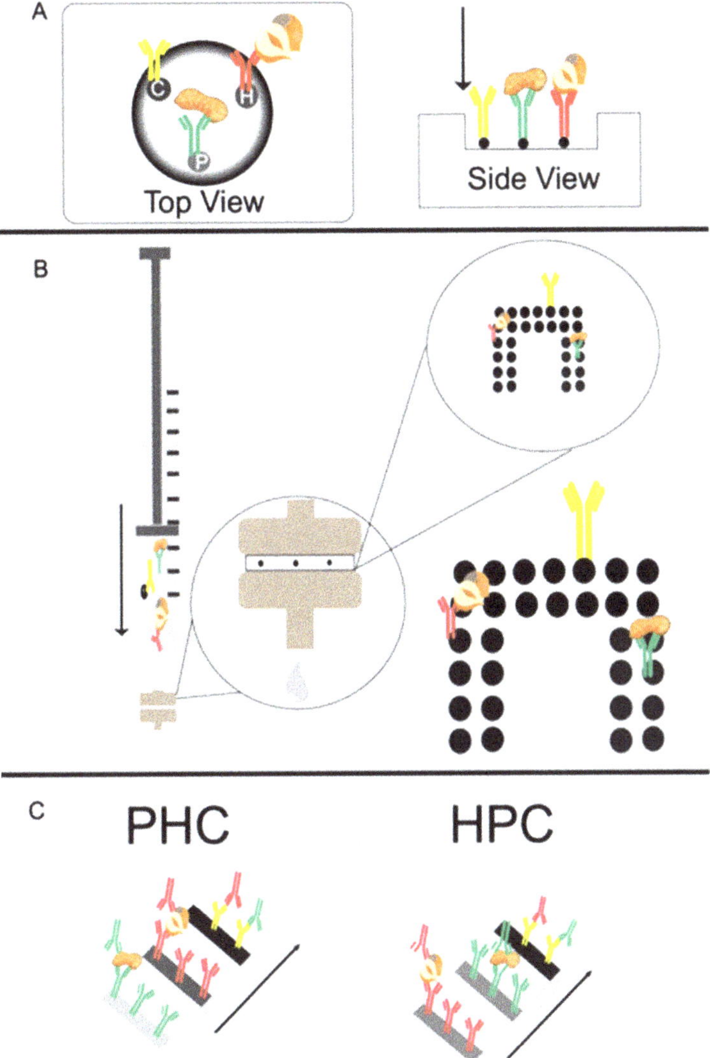

Figure 1. Schematic representation (not to scale) of the three flow assay formats developed. Arrows depict the flow direction and C is the control antibody (goat anti-mouse), H is the anti-hazelnut antibody and P is the anti-peanut antibody. Total hazelnut protein (THP) is indicated by the hazelnut graphic and total peanut protein (TPP) is indicated by the peanut graphic. (**A**) The passive flow assay in top-view and side-view. (**B**) The active format flow-through assay, where the syringe filter holder is enlarged, and the membrane is further enlarged to show the biofunctionalized area. (**C**) Both lateral flow immunoassay geometries as defined by the order in which sample will encounter the test and control lines: Peanut, hazelnut, control (PHC) and hazelnut, peanut, control (HPC).

2.5. Multiplex Active Flow-through

A schematic representation of the active flow-through assay is shown in Figure 1B. First, the most appropriate assays parameters were established including membrane type, pore size, antibody concentration for dispensing and assay conditions.

2.5.1. Simplified Multiplex Flow-through

Allergen-specific antibody solutions (0.5 µL of 1 mg/mL mAb solution) and control antibody solution were manually dispensed by lightly touching the tip of the pipette to the membrane onto 0.2 or 0.45 µm pore size unbacked NC or 0.45 µm unbacked nylon membranes. The membranes were dried for 45 min and then the membranes were placed in 13 mm syringe filter holders and attached to the 10 mL syringe. The assays were performed by manually and sequentially injecting 500 µL sample (concentration series 100–0.1 ppm total protein extract diluted in RB), 1 µL of each CNP-mAb and another 300 µL of RB as a washing step. In this context, sequentially refers to the sequential loading of the syringe with sample with the CNP-mAbs on top of the sample; these were then pushed through by moving the plunger downwards in a single movement, followed by a final washing step with RB. The membranes were then removed from the filter holder, dried for 5 min, read with the naked eye and an image was acquired with a smartphone camera.

2.5.2. Multiplex Flow-through Iterative Optimization

To establish the optimum active flow-through conditions, a number of alternative assay steps were explored. The experiments aimed to reduce background staining, to improve the signal-to-noise ratio and to improve the assay sensitivity.

2.5.3. Volume Optimization

Different sample and reagent volumes were tested to determine the optimum conditions for flow-through operation. Flow-through assays require larger sample volumes compared with LFIA due to reduced contact time between analyte and capture antibodies [42].

When using sample volumes of less than 500 µL, it was necessary to first 'pre-wet' the membrane with running buffer to ensure that the entire surface would be wetted. Initially, membranes were tested using 500 µL RB, followed by a 300 or 500 µL sample and 0.5 µL of each of the CNP-mAbs solutions followed by 500 µL RB as a washing step. In subsequent experiments, the volume of the CNP-mAb solution was increased to 1 µL for each CNP-mAb to maximize the signal intensity. Finally, experiments were performed using 1 mL of sample, with 1 µL of each CNP-mAb solution dispensed on top of the sample, followed by 500 µL RB.

2.5.4. Pre-Mix Method

The assays were tested by pre-mixing the running buffer and CNP-labeled secondary mAbs with sample and injecting the mixture simultaneously. In this approach, 1 mL of sample, 1 mL of RB and 1 µL of each CNP-mAb were injected across the membrane, effectively causing an additional 50% dilution to the sample, when compared to the sequential method described above. The holder was then dismantled, and the membrane dried for 5 min before visual inspection.

2.5.5. Filter Approach

To improve the uniform wetting of the membrane and reduce the background staining caused by the CNPs, a filter approach was tested. In this method, a 0.45 µm NC filter was placed on top of the functionalized membrane before carrying out the assay sequentially. Following the final wash step, the device was dissembled, the 0.45 µm filter carefully removed and disposed of and the membrane dried for 5 min before visual inspection.

2.5.6. Aspiration Approach

To ensure sufficient wetting of the membrane, and to increase the contact time of the sample and the capture antibodies, an iterative aspiration approach was applied. In this way, when sequentially injecting the sample and CNP-mAbs, the plunger of the syringe was pumped up and down, 1, 5 or 10 times. With the increasing number of aspirations, the flux of the analyte past the membrane, and thus past the immobilized antibodies, was increased. After the final aspiration, the RB was flowed through as a washing step, the device was disassembled, and the membrane dried for 5 min before visual inspection and photographing with a smartphone camera.

2.5.7. Multiplex Array Layout

The flow-through array was spotted using the XYZ 3060 BioDot Dispense Platform (Irving, CA, USA). The array was composed of 14 (2 × 7 array) control spots (0.25 mg/mL) and with each analyte having 12 (2 × 6 array) spots (0.25 mg/mL), with a drop size of 100 nL and an offset of 1 mm between each dot (see Figure 1B). The membranes were left to dry overnight prior to testing.

2.5.8. Optimized Active Flow-through Operation Protocol

A 0.45 μm NC filter, acting as a vertical flow diffuser, was placed on top of the biofunctionalized membrane. The filter and membrane were then placed, biofunctionalized side up, into the syringe filter holder. A polytetrafluorothylene (PTFE) gasket was placed on top of the membrane to seal the fluid pathway, giving the assay an actual flow path of 10 mm. The syringe holder was then attached to a 10 mL Luer-Lock™ syringe. The assay was performed sequentially as described in Section 2.5.1. First, 1 mL of sample topped with 1 μL of each CNP-mAb solution was aspirated 10 times across the membrane (only THP or only TPP or mixture of both diluted in RB at 100, 10, 1, 0.1 and 0 ppm). Following this, 500 μL RB, as a washing buffer, was flowed through the membrane. Finally, the syringe filter holder was disassembled, and the membrane removed and placed on an absorbent pad for drying. To determine whether the immobilized test antibodies suffered from non-specific binding towards the other target, the assays were tested using just THP or just TPP extract spiked into RB.

Blank buffer measurements were performed 10 times to test for false positives. The membranes were visually inspected and photographed with a smartphone camera after 5 min. LOD values for visual inspection were established at the lowest concentration that reproducibly yielded a signal that could be observed and distinguished from the background by the naked eye.

2.6. Multiplex Lateral Flow Immunoassay

Lateral flow immunoassays were manufactured using NC (flow rate of 140 s/4 cm) cut to approximately 4 cm length. The NC membrane was secured on a plastic backing, with 4.5 cm of absorbent pad overlapping one end of the NC. Two different test line configurations (as depicted in Figure 1C) were designed and produced using the XYZ BioDot dispensing platform. The first configuration had the control line (0.25 mg/mL) dispensed at 10 mm from the absorbent pad, the hazelnut line (0.25 mg/mL) at 5 mm from the control line and the peanut line at 5 mm from the hazelnut line, with 10 mm of blank membrane at the bottom of the strip, hereafter referred to as PHC. The second arrangement had the control line at 10 mm from the absorbent pad, the peanut line at 5 mm from the control line and the hazelnut line at 7 mm from the peanut line with 8 mm of blank membrane at the bottom of the strip, hereafter referred to as HPC.

Multiplex LFIA Operation Protocol

Firstly, the multiplex LFIAs were tested for non-specific binding by testing 10 × each of the LFIAs in blank running buffer (RB). The LFIAs were inserted into individual microwells of a 96-well plate containing 1 μL of each of the CNP-mAbs and 100 μL of RB (blank). The strips were left to run for 5 min. Next, the LFIAs were tested for specificity by testing in either just THP or TPP extract spiked into RB. LFIAs were placed into the individual microwells of a 96-well plate containing either just THP or TPP (1 μL) spiked into RB, in decreasing concentration with RB (99 μL) and 1 μL of each carbon-labeled mAb. The strips were left to run for 5 min before photographing with a smartphone camera. Finally, the assays were tested using the same conditions in decreasing concentrations (100, 10, 1, 0.5, 0.1 ppm) of both THP and TPP spiked in RB (in triplicate). Calibration series were tested with both formats of the LFIA using (i) 1 μL of sample (diluted in RB) and 99 μL of RB (hereafter, 1:99, sample: RB), (ii) 25 μL of sample (diluted in RB) and 75 μL of RB (hereafter, 25:75, sample: RB), and (iii) 75 μL sample (diluted in RB) and 25 μL of RB (hereafter, 75:25, sample: RB). The 75:25 sample: RB experiments were specifically designed to trigger the hook-effect to determine when the sample volume becomes the limiting factor.

The membranes were visually inspected and photographed with a smartphone camera after running for 5 min. LOD values for visual inspection were established at the lowest concentration that reproducibly yielded a signal that could be observed and distinguished from the background by the naked eye.

2.7. Smartphone Readout and Data Analysis

Smartphone photographs were acquired using Open Camera (version 4.0.3) and analyzed using a Huawei P20 smartphone (Huawei Technologies, Shenzen, China) according to the method developed by Ross et al. [27] using two freely downloadable apps from the Google Play Store. The red, green, blue (RGB) values were obtained for test regions of assays using the RGB Color Detector (version 1.0.58). Using the crosshair function in the app, test dots on the flow-through membrane or three distinct regions on the test line of the LFIA were selected and the color values were averaged and recorded. Background measurements were also made above and below the test areas to determine an overall background level for subtraction from results. Alternatively, results were normalized by dividing the value of each test region by the corresponding control region, as has been performed in literature [35,47,48]. Using 'Nix Pro Color' (version 1.31), the RGB values were converted to luminosity, A, B (LAB) values; a device-independent color space that more accurately represents how humans interpret color intensity.

Additionally, to show the device-independent nature of LAB measurements, the optimized assays were also analyzed using a Google Pixel 2 XL smartphone (Google, Mountain View, CA, USA). The obtained values were used to plot calibration curves for L (luminosity) of the LAB values as a function of allergen concentrations spiked into RB, using Microsoft Excel. LOD values were obtained from these calibration curves by visual evaluation.

2.8. Matrix Experiments and Validation

To validate the assays, they were also tested in spiked food matrices. All assays were tested in a decreasing concentration of THP and TPP, spiked directly into a blank biscuit matrix extract to determine the matrix effects. Additionally, the optimized LFIA (PHC) was more extensively validated by testing in 20 truly different blank matrix extracts. In this way, LFIAs were placed in individual microwells containing 25 μL blank matrix extract (n = 20) and 75 μL RB and left to run for 10 min to determine whether any false positives occurred. Additionally, 1 ppm of THP and TPP was spiked into the 20 different blank matrix extracts (1 μL of 1000 ppm THP and TPP sample into 999 μL (*v/v*) blank matrix extract) and the LFIAs were tested using both 25 μL spiked matrix plus 75 μL RB and 1 μL spiked matrix extract plus 99 μL RB. Assays were left to develop for 10 min. Finally, the optimized LFIAs were also tested in blank flour matrix extract and spiked peanut flour matrix extract in both 25:75 and 1:99 dilutions in RB.

3. Results and Discussion

3.1. Multiplex Passive Flow-through Assay

An overview of conditions, quantitative and qualitative results for spiked buffer experiments for the passive flow-through assay, can be found in Table 1. The visual limit of detection (LOD) for the passive flow-through was established by testing in decreasing concentrations of THP and TPP extracts spiked in RB. The visual LODs were determined as 0.1 ppm and 1 ppm and smartphone LODs 1 and 10 ppm for hazelnut and peanut, respectively ($n = 3$), whereas no visible spot was obtained for blanks (see Table 1 and Supplementary Information, Figure S1A). Following the addition of the CNP-mAbs to the passive flow-through assay, the positive spots appeared within 5 s, a detection speed which is unparalleled by LFIA. Even when using the high-speed LFIA described in [30] the appearance of the positive result took 30 s, due to MTL limitations of the solution that needs to wick through the membrane before reaching test lines. Three drops of RB were added to the flow-through assay to wash the unbound CNPs from the membrane. Using dropper bottles with pre-defined drop volumes for the delivery of RB makes the assay easy to perform and means that pipettes are unnecessary. A further benefit is that the result can be directly read through the window of the cassette by the naked eye without having to disassemble the device. However, when recording a smartphone image of the membranes, these do need to be removed from the plastic cassette to avoid shadowing. Despite the washing step, the membranes had variable background staining, which made it impossible to obtain calibration curves from the images acquired with a smartphone. The reason for the appearance of background staining probably lies with the polydispersity of the CNP, which can form aggregates of several hundred nm, which are too large to be flowed through the pores. A drawback of this specific passive flow assay format is the lack of freedom in geometric assay design as bio-reagents required manual spotting by pipette. However, such a limitation could be easily overcome by biofunctionalization of the membranes before having them cut to the factory-made circular size.

Table 1. Comparison of optimized Flow-through and Lateral Flow parameters (RB *).

Parameter	Passive Flow-through	Active Flow-through	PHC **	HPC **
Visual/Smartphone LOD (ppm); **Hazelnut (h)** **Peanut (p)**	h: 0.1/1 p: 1/10	h: 0.05/0.5 p: 0.05/0.5	h: 0.1/0.5 p: 0.5/0.5	h: 1/5 p: 5/10
Working Range (ppm)	1000–0.1	1000–0.05	100–0.1	10,000–0.1
Assay Duration (total assay time, incl. drying)	5 min	10 min	5 min	5 min
Time until result appearance	5 s	5 s	30 s–1 min	1 min
Sample extracted volume (μL)	50	1000	25	1
Flexibility of multiplexing	Low—requires manual dispensing bioreagents	High—Printing nL/μL size dots or multi-line	Medium—test line configuration and positioning of antibodies has an influence.	Medium—test line configuration and positioning of antibodies has an influence.
Non-Expert Ease of Use	Easy	Challenging	Easy	Easy
False positives in blank RB (*n* = number of tested samples within assay working range)	Y (*n* = 3)	Y (*n* = 10)	N (*n* = 10)	N (*n* = 10)
False negatives in spiked RB (*n* = number of tested samples within assay working range)	N (*n* = 3)	N (*n* = 21)	N (*n* = 24)	N (*n* = 18)
Equipment used	Assay cassettes, dropper bottle, pipette	10 mL syringe, syringe filter holder, assay membrane, additional filter, pipette, waste beaker	LFIA, pipettes, microwell plate	LFIA, pipettes, microwell plate
Waste	High plastic consumption (cassettes)	High plastic consumption (syringes) + need for disposal of high volumes liquid waste	Nitrocellulose strips and well plate + disposal of small volume liquid waste	Nitrocellulose strips and well plate + disposal of small volume liquid waste

* All measurements were made using total hazelnut protein and total peanut protein (THP and TPP) spiked into running buffer (RB). ** Where the peanut, hazelnut, control geometry is defined by PHC and the hazelnut, peanut, control geometry is defined by HPC.

3.2. Multiplex Active Flow-through

An overview of conditions, quantitative and qualitative results for spiked buffer experiments for the active flow-through assay can be found in Table 1. The assays using the 0.45 µm pore size nylon and NC membranes were ineffective, and no spots (including control spots) appeared on these membranes. This can be attributed to 0.45 µm being too large a pore size and the majority of the analyte and labeled antibodies passing through the membrane, which is confirmed by the dark coloration of the waste liquid when using this assay membrane. Therefore, the 0.2 µm pore size NC membrane was determined to be the most suitable for this application.

During the optimization steps, active flow-through assays were tested using 0.5 µL of each CNP-mAb solution, but this only yielded faint detection spots. In subsequent experiments the volume of the CNP-mAb solution was increased to 1 µL of each CNP-mAb which improved the readability. Additionally, volumes of 500 µL and 1 mL of sample were tested, with the sensitivity improving with the increased sample volume, without the appearance of a hook-effect, even at high concentrations. Although in this manually spotted initial format, LODs of 0.5 and 0.1 ppm could be reached for peanut and hazelnut (see Supplementary Information, Figure S1B), respectively, false positives were also detected when testing the assays in a blank sample (1 in 5 false positives). Using a pre-mix approach did improve the overall user-friendliness of the assay, as the operator only needed to pass the liquid containing the sample, CNP-labeled mAbs and RB through once without the necessity of removing and reinserting the plunger, but this method consistently resulted in false positives in the blank samples. Contrastingly, using the sequential method increased the difficulty of the assay, but prevented false positives owing to the washing step at the end. The addition of a 0.45 µm NC filter on top of the biofunctionalized membrane increased the (smartphone) readability of the assay. Besides filtering the larger sized CNPs, reducing the level of background staining, the filter also acted as a flow diffuser. In this way, uniform wettability of the membrane was achieved, resulting in better reproducibility compared to when it was performed without the filter. Although the filter improved the readability of the membranes, it also further complicated the user-friendliness of the method, as it needed to be carefully removed from the biofunctionalized membrane before the results could be read.

The sensitivity of the assay was improved by increasing the number of sample aspirations across the membrane (see Supplementary Information, Figure S2). Flow-through assays are subject to unidirectional flow and require capture antibodies with rapid association rates in order to achieve binding or require extended sample/reagent incubation times [48]. By increasing the number of sample aspirations, the flux of the CNP-mAb-analyte complex past the immobilized antibodies, and the potential of binding, is increased. Of all the tested parameters the most appropriate assay conditions were determined to be a 0.45 µm filter on top a 0.22 µm NC membrane biofunctionalized with 0.25 mg/mL control and test spots and aspirating 1 mL of sample with 1 µL of CNP-mAb solution 10 times back and forth through the membrane. Subsequently, 500 µL of RB was injected as a washing step. Although these conditions allowed for the assay to reach very low LODs, they also meant that this method generated a high volume of chemical waste (1.5 mL), which needs to be safely disposed of.

When testing active flow-through membranes in decreasing concentrations of THP and TPP spiked into RB, visual LODs of 0.05 ppm ($n = 3$) could be reached for both targets, an LOD which is so far un-met by commercially available allergen assays [8]. This LOD is less obvious from the smartphone image (LODs of 0.5 ppm for both THP and TTP) compared with reading by naked eye (see Figure 2). Therefore, eye symbols are inserted in Figure 2 to designate the lowest concentration that could still be read visually. Despite the active flow-through approach reaching lower LODs than the passive flow-through assay, the assay was more complicated to perform and used a far greater sample volume.

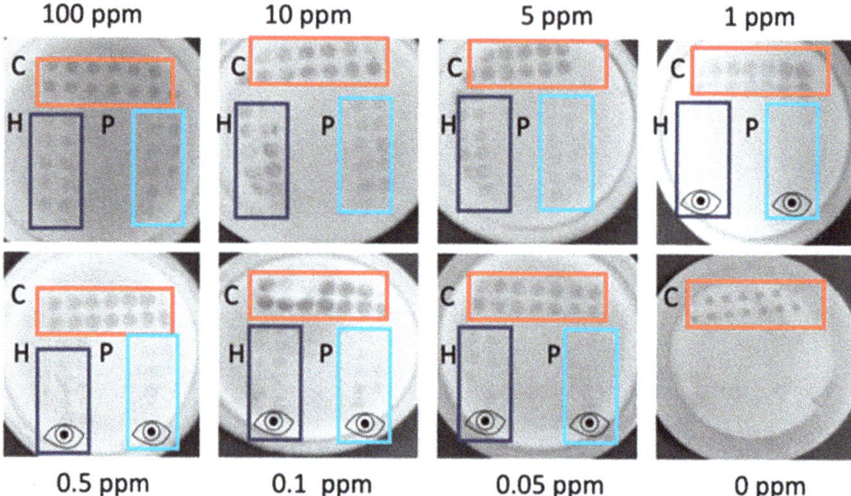

Figure 2. Active flow-through assay calibration range. Assays were tested in decreasing concentrations (100–0.05 ppm) of Total Hazelnut Protein (THP), Total Peanut Protein (TPP) spiked into Running Buffer (RB) and in blank RB. The control region is indicated by C and outlined in red, the hazelnut region by H and outlined in dark blue and the peanut region by P and outlined in light blue. There is an evident decrease in test dot intensity as the concentration of total protein in the sample decreases. The eye icon is used to indicate test regions that are visible to the naked eye but more difficult to read in the smartphone image. The visual limit of detection is established at 0.05 ppm for both analytes.

3.3. Multiplex Lateral Flow Immunoassay

An overview of conditions, quantitative and qualitative results for spiked buffer experiments for the LFIAs can be found in Table 1. The LFIAs were both able to achieve single analyte detection and a true blank result every time (0% false positives at 0 ppm; $n = 10$). When testing PHC with 1 µL of sample, 1 µL of each CNP-mAb and 99 µL of RB, visual LODs of 1 and 5 ppm were achieved by the naked eye (see Figure 3A) for hazelnut and peanut, respectively, with a clear decrease in intensity in the test line with decreasing concentration of the sample. When the LFIAs have a low signal intensity, the naked eye is still superior at distinguishing between a positive or negative signal, and the lower visual LODs are indicated by the eye icon in Figure 3. However, these visual readings are performed by a trained person, and the distinction between signal and no signal at the lowest concentrations is not trivial. In comparison, when the same anti-hazelnut antibody was applied in a single-plex LFIA, an LOD of 0.1 ppm in spiked buffer was reached, which suggests that having an additional test line on the LFIA can compromise the overall sensitivity [30]. Still, the multiplex LODs are in accordance with commercially available allergen single-plex LFIAs, which report LODs within this range. However, lack of standardized, certified reference materials in the allergen industry means that each reported assay is developed using antibodies specific to different allergenic components (total soluble protein vs. allergen-specific proteins) and tested and validated using different analytes [9,45], thus underlining that true comparisons can only be made when bioreagents and samples are kept constant, as in this research. To optimize the multiplex LFIA and improve the LOD, the sample volume was increased to 25 µL (diluted in RB) in 75 µL RB. By increasing the sample volume to 25 µL (thus concentrating the sample 25 × compared with the 1 µL sample volume) LODs of 0.1 and 0.5 ppm for hazelnut and peanut were reached respectively (see Figure 3B).

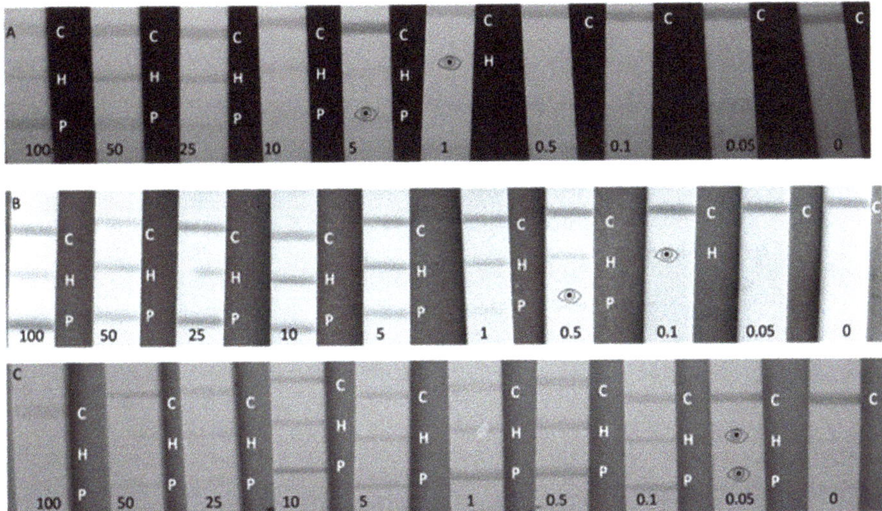

Figure 3. Calibration range (100–0.05 ppm) of Total Hazelnut Protein (THP), Total Peanut Protein (TPP) spiked into Running Buffer (RB) and blank RB, where the control line is indicated by C, the hazelnut test line by an H and the peanut test line by a P. A positive result can be still read with the naked eye, but is difficult to see in the smartphone image, thus an eye icon has been used to indicate the visual LOD. (**A**) Peanut, Hazelnut, Control (PHC) line configuration using 1 µL of spiked sample and 99 µL RB. (**B**) PHC using 25 µL of spiked sample and 75 µL RB. (**C**) PHC using 75 µL of spiked sample and 25 µL RB.

Despite the assay sensitivity improving with the increased sample volume, with these conditions at concentrations of 100 ppm and higher, a reduction of the intensity of the upper line (hazelnut) could be observed, as has been witnessed by Galan-Malo et al. [24]. Although this was not considered a false negative, as three distinct lines were still clearly visible, it did warrant further exploration into the extent of the hook-effect in more concentrated samples.

To further investigate the extent of the hook-effect and its potential to limit the upper dynamic range of the LFIA assay, the PHC format was also tested in 75 µL of sample extract diluted with 25 µL RB (see Figure 3C). These conditions resulted in a more pronounced hook-effect with LFIAs tested at 1000 ppm appearing to be false negatives, and at 100–50 ppm exhibiting decreased test line signals. As well as just testing high analyte concentrations, it is important to test different sample-to-RB ratios, as increasing sample volume has a noteworthy influence on the appearance of the hook-effect. In order to avoid the hook-effect it is imperative to use the correct volume of diluted sample. Despite this, PHC in the 75:25 conditions did achieve a lower LOD of 0.05 ppm for both analytes in RB. Therefore, PHC could still be used with 75:25 conditions for testing trace allergen levels, so long as the sample is also tested in the 1:99 and 25:75 conditions to ensure no false negatives arise at high concentrations. The optimum conditions from PHC were determined to be 25:75. When testing HPC in the 1:99 conditions, LODs of 5 and 1 ppm (see Supplementary Information, Figure S3) were reached for peanut and hazelnut, respectively, with the LODs decreasing to 1 and 0.1 with the 25:75 arrangement. But for HPC, the hook-effect was greater in 25:75 compared with PHC with concentrations of 100 and 50 ppm experiencing reduced intensity on both the control and the peanut lines, complicating quantitative analysis. The larger hook-effect in this configuration could be because the upstream (hazelnut) test line comes into contact with the sample first, and this mAb has a rapid association rate and high affinity for THP, and so it becomes quickly saturated [30].

So, the optimum condition for HPC was the 1:99 protocol, although this was significantly less sensitive compared with the optimized PHC assay. For this reason, PHC was determined to be the optimum test line configuration with the best working conditions being 25:75 in the working range of 100–0.1 ppm. Therefore, PHC was used for further smartphone quantification and validation experiments.

3.4. Smartphone Readout and Analysis

Smartphones are ever-increasing in popularity for analyzing colorimetric assays. Most often, smartphone analysis is based on specific apps which relate a particular color intensity to a certain concentration of analyte. In the absence of a specific app, it has been shown by Ross et al. [30] that it is possible to use freely downloadable apps from the Google Play Store to analyze endpoint, smartphone image color intensity values. By converting RGB values to LAB values, luminosity or intensity can be plotted as a function of concentration in a calibration curve. In sandwich immunoassay formats with CNP labels, a higher L value corresponds to a lower analyte concentration. As LAB color space is device-independent, the same results can be potentially achieved using different smartphone models. For analysis of PHC and HPC (in triplicate) the normalization of the (L)LAB values was carried out by dividing the L values of the test lines by the L values of the control lines. The method of dividing the test line response by the control line response (T/C ratio) is a technique commonly used for the quantification of sandwich LFIAs [35,47–49]. The results for PHC can be found in Figure 4, and the HPC smartphone calibration curve can be found in Figure S4 in Supplementary Information.

Two smartphone models were used for the device independent LAB analysis of PHC assays (in RB in triplicate), as can be seen in Figure 4 where A, C and E show the curves for THP in 1:99, 25:75 and 75:25 (sample: RB) and B, D and F show the curves for TPP in 1:99, 25:75 and 75:25 (sample: RB). A higher normalized L value was obtained for hazelnut at 25–100 ppm using the 25:75 conditions, as can be seen in Figure 4C. Comparatively, peanut did not appear to be subject to the hook-effect under 25:75. Using 75:25 conditions (see Figure 4E), concentrations of 50 and 100 ppm resulted in a higher normalized L value for hazelnut (i.e., weaker signal). Furthermore, under these conditions the hazelnut T/C ratio for 10 ppm and 25 ppm gave the same normalized L value, highlighting that the hook-effect was still evident, even at these lower concentrations. Comparatively, peanut in 75:25 (see Figure 4F) gave higher normalized L values at concentrations of 25–100 ppm, again indicating with increasing sample volume and concentration the likelihood of the hook-effect being increased. The only crucial variation between the two smartphone measurements using the different models was obtained for the peanut line using 75:25 at 0.1 ppm (see Figure 4F). However, this is the smartphone LOD, and detection spots were already more difficult to read. As well as this, the current method relies on manually selecting regions of interest on the control and test lines, rather than being able to read the values across the whole line. Therefore, please note that the results also include any errors due to not selecting the exact same regions, and this can also cause variations in the obtained color values.

Additionally, to compare different smartphone quantification methods, all smartphone readable assays were also analyzed by making a background subtraction as can be seen in Supplementary Information (see Figures S5 and S6). However, when analyzing the LFIAs in this way the differences in background readings, due to discrepancies in lighting conditions caused by recording an image of the entire calibration range simultaneously under ambient lighting conditions, meant that a simple background subtraction was insufficient. However, for active flow-through assays the background subtraction was found to be the most effective analysis method (see Supplementary Information, Figure S6A), whereas the T/C method resulted in larger standard deviations (see Supplementary Information, Figure S6B). This could be attributed to the membranes being photographed independently, so the small membranes were subject to the same ambient lighting conditions and did not have such variable background readings. By using two data processing methods it is evident that the selected data processing method plays a crucial role for the quality of the semi-quantitative information that can be obtained from raw results.

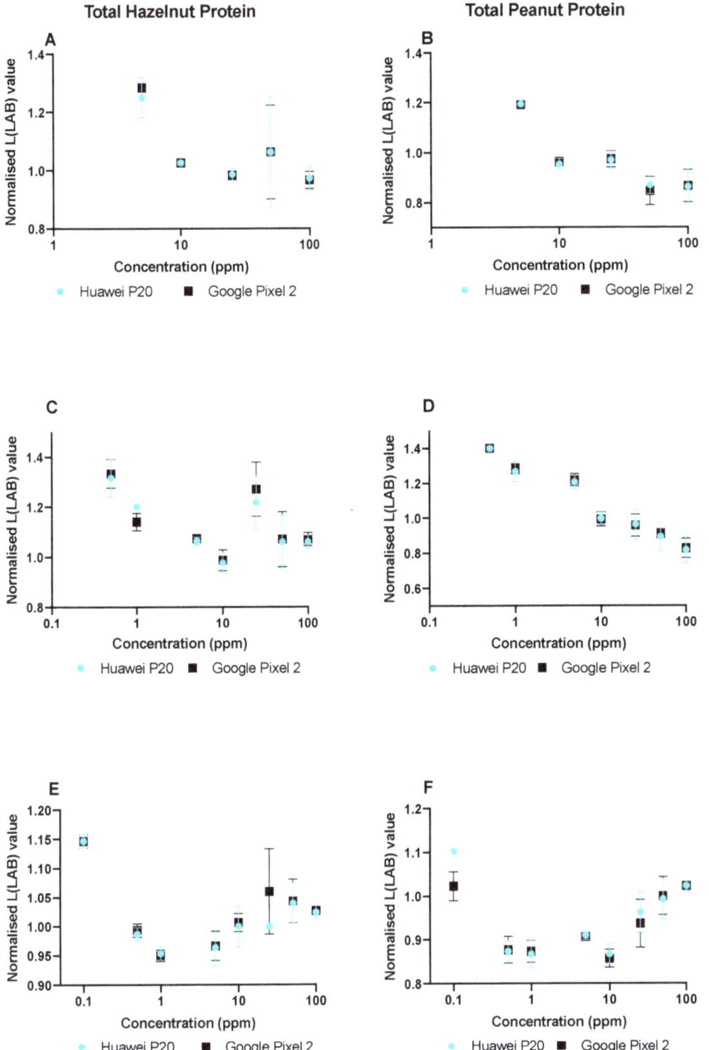

Figure 4. Smartphone calibration curves for the normalized (L) LAB values of the test lines of a Peanut Hazelnut Control (PHC) assay as a function of the concentration of Total Hazelnut Protein (THP), and Total Peanut Protein (TPP) (100–0.1 ppm) tested using two different smartphone models. All calibration ranges were performed in triplicate in spiked Running Buffer (RB). All L(LAB) values have been normalized by dividing the test line values by the control line values. (**A**) Hazelnut tested in 1 μL of sample in 99 μL of running buffer (RB) (**B**) Peanut tested in 1 μL of sample in 99 μL of RB. (**C**) Hazelnut tested in 25 μL sample in 75 μL of RB. (**D**) Peanut tested in 25 μL sample in 75 μL of RB. (**E**) Hazelnut tested in 75 μL of sample in 25 μL of RB. (**F**) Peanut tested in 75 μL of sample in 25 μL of RB. Error bars show standard deviation (SD) from triplicate measurements.

3.5. Matrix Experiments and Validation

To determine their applicability to real life samples, the assays were tested using THP and TPP spiked into blank biscuit matrix extracts. The passive flow-through format was able to achieve visual LODs of 5 and 1 ppm for peanut and hazelnut. These LODs are higher than previously observed in spiked buffer experiments, showing that the matrix extract did have some influence on the detection of the analytes. When testing in this way, the passive flow membranes had greater background staining compared with in spiked buffer experiments. This can be attributed to the overall reduction of reagents, BSA and tween-20 in the assay buffer, as the sample was spiked into a matrix extract rather than into the RB.

In comparison, the active flow-through membranes did not suffer with increased background staining due to the use of the additional filter on top of the membrane and subsequent washing step. The active-flow assay reached visual LODs of 0.5 and 1 ppm for THP and TPP in spiked matrix extract, however the intensity of the detection spots was fainter compared with spiked buffer samples because of the reduction of buffer reagents responsible for good flow.

Therefore, whilst visual readout was possible, the construction of calibration curves based on smartphone images could not be achieved.

PHC was tested in both 25:75 and 1:99 of spiked matrix in RB to determine the visual LOD in matrix extract, as can be seen in Supplementary Information Figure S7. When using 25 μL sample (THP and TPP spiked into matrix extract) and 75 μL RB a LOD of 0.5 ppm could be reached for both analytes (see Supplementary Information, Figure S7A). At higher concentrations (100 ppm +) there was decreased intensity for the hazelnut line. This can be attributed to the hook-effect. For the spiked matrix extract experiments, the PHC assays were run for 10 min, due to the reduction of reagents BSA and tween-20 from spiking sample into matrix extract rather than RB, affecting the flow of the sample. Additionally, PHC was tested in 1 μL of spiked matrix extract:99 μL of RB (see Supplementary Information, Figure S7B). Visual LODs of 10 and 5 ppm were reached for peanut and hazelnut, respectively. The PHC assay was fully validated using 25:75 conditions by evaluating 20 truly different blank matrices and determining that no false positives occurred. Additionally, the 20 blank matrices were spiked with 1 ppm THP and TPP. In the absence of agreed regulatory levels for food allergens, a screening target concentration (STC), based on VITAL 2.0 levels of 1 ppm, was selected [8,50]. The LFIAs were able to detect the allergens with both visual and smartphone readout at 1 ppm in all 20 samples, as can be seen in Figure 5 and as is summarized in Table 2. The excellent reproducibility at the STC level clearly suggests that a simple device-independent smartphone readout may provide semi-quantitative data.

Table 2. Matrix experiments for the optimized PHC assay, all measurements made in spiked matrix extract.

Parameter	PHC * (Matrix Extract)
LOD	0.5 ppm both analytes
Working range	100–0.5 ppm
Assay duration (total assay time incl. drying)	10 min
Time to result	1.5–2 min
Sample volume	25 μL
Reproducibility ** ($n = 20$)	Hazelnut: 2.5% Peanut: 3.4%
False positives ($n = 20$)	0
False negatives ($n = 20$)	0

* PHC = Peanut, hazelnut, control geometry lateral flow immunoassay. ** Reproducibility defined as Relative Standard Deviation (RSD) × 100% 1 ppm of Total Hazelnut Protein (THP), Total Peanut Protein (TPP) spiked into blank biscuit matrix extract ($n = 20$). Data based on normalized L (LAB) values.

Finally, to confirm the capability of the optimized LFIA in detecting allergens in raw ingredients, blank flour and peanut-spiked flour samples were briefly tested. The LFIAs correctly did not detect either of the allergens in the blank flour ($n = 4$). Furthermore, PHC specifically detected only peanut in the peanut-spiked flour (n = 4) with no false hazelnut positives being observed. The detection of peanut was not adversely affected by using the accelerated 30 min extraction procedure for the spiked flour. Further developments should include simplified and faster extraction methods.

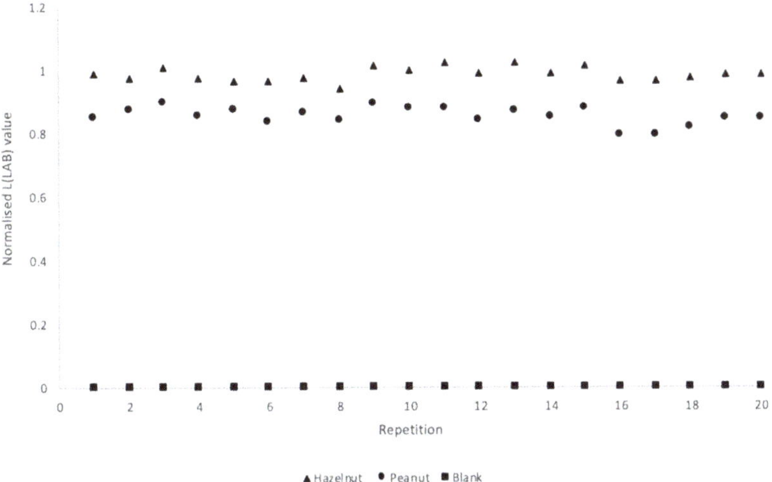

Figure 5. Smartphone validation of Peanut Hazelnut Control (PHC) assay using 20 truly different blank biscuit samples (square markers) and 20 truly different biscuit samples spiked at the screening target concentration of 1 ppm Total Hazelnut Protein (THP), and total peanut protein (TPP). Normalized L (LAB) values were obtained by dividing the test line response by the corresponding control line response.

4. Conclusions

Quick and accurate detection of food allergens is of critical importance for food safety; it is particularly relevant if such testing procedures can be easily performed by the consumer, and therefore, there is an evident requirement for simple and robust testing procedures. Two formats of multiplex flow-through immunoassays have been developed and compared with two test line configurations of LFIA, all developed using the same bioreagents and against the same targets in order to allow a true comparison.

Two recent review papers have extensively outlined commercially-available and proof-of-concept single-plex and multiplex allergen immunoassays and biosensors, and the assays reported in this study have matched or surpassed these previously-reported LODs [9,51]. All the developed multiplex assays were able to detect both analytes in the low ppm range within minutes. It is important to note here that our screening concentrations always related to total protein extracts from either peanuts or hazelnut, and therefore, the concentration of specific allergenic proteins is expected to be even lower than the reported values. This in turn means that the reported LODs are underestimating the true sensitivity of the immunoassays in this work. The passive flow-through format offered a way to rapidly develop a fast flow-through assay. However, this specific format was limited by the need to manually biofunctionalize the membranes, limiting their reproducibility. The active flow-through assay could achieve very low limits of detection with no false negatives when following the optimization steps. However, it is these optimization steps that made the assay more complicated to perform for a non-expert user such as a consumer. In future versions, the use of a mechanical pump could improve the user-friendliness, although this would introduce an additional and costly element into the procedure, limiting the portability of the assay. It should be reiterated that the assays within this study were performed by a trained scientist, and the active flow-through method is not recommended for untrained users. In comparison, the LFIAs, when using the optimized assay conditions for each configuration, resulted in no false positives. However, outside of working conditions, both configurations of LFIA did experience a hook-effect at high concentrations, a phenomenon commonly encountered in sandwich LFIA, where a falsely low signal occurs at high analyte concentrations. As the hook-effect is concentration-dependent, it can be avoided/limited by assay optimization.

To demonstrate their applicability to real life bakery products and raw ingredients, all assays were tested in decreasing concentrations of analyte spiked into the matrix extract. Additionally, the PHC assay was validated as a screening method in spiked matrix extract, blank matrix extract ($n = 20$) and incurred spiked flour, proving its capability of detecting the target even in complex matrices. The majority of commercially-available allergen detection LFIA test kits can detect a single analyte at 1–10 ppm [9,51]. Comparatively, PHC was able to detect both analytes at 0.5 ppm of THP and TPP spiked into a blank biscuit matrix extract, affirming its place as one of the most sensitive allergen LFIAs. This LOD was in agreement with the LOD using the same hazelnut antibody in a previously-reported single-plex assay [30]. Finally, all assays were (semi-)quantified by smartphone readout. At this stage no additional external equipment was used for the image recording, so the LFIA membranes were subject to ambient lighting conditions. To compensate for the lighting conditions a normalization factor (T/C ratio) was applied. By using device-independent (L)LAB values, it was possible to obtain comparable results using two distinct smartphone models. The ability to use different smartphone models for reading the same assays is a characteristic that is highly desirable, but not often reported, within smartphone analysis. In future developments, researchers should focus on improving the ease of use of these assays by integrating sample preparation, limiting the user interaction with the assay, as well as by developing a consumer-friendly app as a user interface which can directly analyze data with minimal user input.

Supplementary Materials: The following are available online at http://www.mdpi.com/2079-6374/9/4/143/s1, Figure S1: Calibration range for multiplex flow-through assays using passive and active flow. Figure S2: Calibration range for multiplex flow-through assay optimization: Sample aspirations. Figure S3: Calibration range for lateral flow immunoassay with test line configuration: Hazelnut, peanut, control (HPC). Figure S4: Calibration curve for smartphone analysis of hazelnut, peanut, control (HPC) lateral flow immunoassay. Figure S5: Calibration curve for smartphone analysis of lateral flow immunoassays using background subtraction. Figure S6: Calibration curve for smartphone analysis of active flow-through immunoassay. Figure S7: Optimized lateral flow immunoassay calibration curve in spiked matrix extract. Table S1: Ingredient and allergen information for the 20 varieties of biscuit used for matrix experiments.

Author Contributions: Conceptualization, G.M.S.R. and G.I.S.; methodology, G.M.S.R.; software, G.M.S.R.; validation, G.M.S.R. formal analysis, G.M.S.R.; investigation, G.M.S.R.; resources, G.M.S.R., G.I.S. & M.W.F.N.; data curation, G.M.S.R.; Writing—Original Draft preparation, G.M.S.R.; Writing—Review and Editing, G.M.S.R., G.I.S. & M.W.F.N.; visualization, G.M.S.R.; supervision, G.I.S. & M.W.F.N.; project administration, M.W.F.N.; funding acquisition, M.W.F.N.

Funding: This project has received funding from the European Union's Horizon 2020 research and innovation program under the Marie Sklodowska-Curie grant agreement No. 720325.

Acknowledgments: The authors would like to thank project partner Barilla for providing test materials for blank and spiked flour samples.

Conflicts of Interest: The authors declare no conflict of interest.

References

1. O'Keefe, A.W.; De Schryver, S.; Mill, J.; Mill, C.; Dery, A.; Ben-Shoshan, M. Diagnosis and management of food allergies: New and emerging options: A systematic review. *J. Asthma Allergy* **2014**, *7*, 141–164. [CrossRef]

2. Weinberger, T.; Sicherer, S. Current perspectives on tree nut allergy: A review. *J. Asthma Allergy* **2018**, *11*, 41–51. [CrossRef]

3. McWilliam, V.; Koplin, J.; Lodge, C.; Tang, M.; Dharmage, S.; Allen, K. The Prevalence of Tree Nut Allergy: A Systematic Review. *Curr. Allergy Asthma Rep.* **2015**, *15*, 1–54. [CrossRef]

4. Maloney, J.M.; Rudengren, M.; Ahlstedt, S.; Bock, S.A.; Sampson, H.A. The use of serum-specific IgE measurements for the diagnosis of peanut, tree nut, and seed allergy. *J. Allergy Clin. Immunol.* **2008**, *122*, 145–151. [CrossRef]

5. European Commission. Directive, EC 2003/89/EC of the European Parliament and of the Council of 10 November 2003 Amending Directive 2000/13/EC as Regards Indication of the Ingredients Present in Foodstuffs. *OJEU* **2003**, *308*.

6. European Union. Regulation (EU) No 1169/2011 of the European Parliament and of the Council of 25 October 2011 on the provision of food information to consumers, amending Regulations (EC) No 1924/2006 and (EC) No 1925/2006 of the European Parliament and of the Council. *OJEU* **2011**, *L304*, 18–63.

7. Allen, K.J.; Turner, P.J.; Pawankar, R.; Taylor, S.; Sicherer, S.; Lack, G.; Rosario, N.; Ebisawa, M.; Wong, G.; Mills, E.N.C.; et al. Precautionary labelling of foods for allergen content: Are we ready for a global framework? *World Allergy Organ. J.* **2014**, *7*, 1–14. [CrossRef]

8. Soon, J.M.; Manning, L. May contain allergen statements: Facilitating or frustrating consumers? *J. Consum. Policy* **2017**, *40*, 447–472. [CrossRef]

9. Ross, G.M.; Bremer, M.G.; Nielen, M.W. Consumer-friendly food allergen detection: Moving towards smartphone-based immunoassays. *Anal. Bioanal. Chem.* **2018**, *410*, 5353–5371. [CrossRef]

10. Choi, J.R.; Yong, K.W.; Choi, J.Y.; Cowie, A.C. Emerging Point-of-care Technologies for Food Safety Analysis. *Sensors* **2019**, *19*, 817. [CrossRef]

11. Zhang, J.; Portela, S.B.; Horrell, J.B.; Leung, A.; Weitmann, D.R.; Artiuch, J.B.; Wilson, S.M.; Cipriani, M.; Slakey, L.K.; Burt, A.M.; et al. An integrated, accurate, rapid, and economical handheld consumer gluten detector. *Food Chem.* **2019**, *275*, 446–456. [CrossRef] [PubMed]

12. Taylor, S.L.; Nordlee, J.A.; Jayasena, S.; Baumert, J.L. Evaluation of a handheld gluten detection device. *J. Food Prot.* **2018**, *81*, 1723–1728. [CrossRef] [PubMed]

13. Alves, R.C.; Barroso, M.F.; González-García, M.B.; Oliveira, M.B.P.; Delerue-Matos, C. New Trends in Food Allergens Detection: Toward Biosensing Strategies. *Crit. Rev. Food Sci. Nutr.* **2016**, *56*, 2304–2319. [CrossRef] [PubMed]

14. Rateni, G.; Dario, P.; Cavallo, F. Smartphone-Based Food Diagnostic Technologies: A Review. *Sensors* **2017**, *17*, 1453. [CrossRef]

15. Lin, H.-Y.; Huang, C.-H.; Park, J.; Pathania, D.; Castro, C.M.; Fasano, A.; Weissleder, R.; Lee, H. Integrated Magneto-Chemical Sensor for On-Site Food Allergen Detection. *ACS Nano* **2017**, *11*, 10062–10069. [CrossRef]

16. Coskun, A.F.; Wong, J.; Khodadadi, D.; Nagi, R.; Tey, A.; Ozcan, A. A personalized food allergen testing platform on a cellphone. *Lab Chip* **2013**, *13*, 636–640. [CrossRef]

17. Zhang, D.; Liu, Q. Biosensors and bioelectronics on smartphone for portable biochemical detection. *Biosens. Bioelectron.* **2016**, *75*, 273–284. [CrossRef]

18. Morón, M.J.; Luque, R.; Casilari, E. On the capability of smartphones to perform as communication gateways in medical wireless personal area networks. *Sensors* **2014**, *14*, 575–594. [CrossRef]

19. Gantelius, J.; Bass, T.; Sjöberg, R.; Nilsson, P.; Andersson-Svahn, H. A lateral flow protein microarray for rapid and sensitive antibody assays. *Int. J. Mol. Sci.* **2011**, *12*, 7748–7759. [CrossRef]

20. Anfossi, L.; Di Nardo, F.; Cavalera, S.; Giovannoli, C.; Baggiani, C. Multiplex Lateral Flow Immunoassay: An Overview of Strategies towards High-throughput Point-of-Need Testing. *Biosensors* **2018**, *9*, 2. [CrossRef]

21. Peng, J.; Wang, Y.; Liu, L.; Kuang, H.; Li, A.; Xu, C. Multiplex lateral flow immunoassay for five antibiotics detection based on gold nanoparticle aggregations. *RSC Adv.* **2016**, *6*, 7798–7805. [CrossRef]

22. Song, S.; Liu, N.; Zhao, Z.; Njumbe Ediage, E.; Wu, S.; Sun, C.; De Saeger, S.; Wu, A. Multiplex Lateral Flow Immunoassay for Mycotoxin Determination. *Anal. Chem.* **2014**, *86*, 4995–5001. [CrossRef]

23. Anfossi, L.; Di Nardo, F.; Russo, A.; Cavalera, S.; Giovannoli, C.; Spano, G.; Baumgartner, S.; Lauter, K.; Baggiani, C. Silver and gold nanoparticles as multi-chromatic lateral flow assay probes for the detection of food allergens. *Anal. Bioanal. Chem.* **2018**. [CrossRef] [PubMed]

24. Galan-Malo, P.; Pellicer, S.; Pérez, M.D.; Sánchez, L.; Razquin, P.; Mata, L. Development of a novel duplex lateral flow test for simultaneous detection of casein and β-lactoglobulin in food. *Food Chem.* **2019**, *293*, 41–48. [CrossRef] [PubMed]

25. Cho, D.G.; Yoo, H.; Lee, H.; Choi, Y.K.; Lee, M.; Ahn, D.J.; Hong, S. High-Speed Lateral Flow Strategy for a Fast Biosensing with an Improved Selectivity and Binding Affinity. *Sensors* **2018**, *18*, 1507. [CrossRef] [PubMed]

26. Zhao, M.; Wang, X.; Nolte, D. Mass-transport limitations in spot-based microarrays. *Biomed Opt. Express* **2010**, *1*, 983–997. [CrossRef] [PubMed]

27. Chen, P.; Gates-Hollingsworth, M.; Pandit, S.; Park, A.; Montgomery, D.; AuCoin, D.; Gu, J.; Zenhausern, F. Paper-based Vertical Flow Immunoassay (VFI) for detection of bio-threat pathogens. *Talanta* **2019**, *191*, 81–88. [CrossRef]

28. Bishop, J.D.; Hsieh, H.V.; Gasperino, D.J.; Weigl, B.H. Sensitivity enhancement in lateral flow assays: A systems perspective. *Lab Chip* **2019**, *19*, 2486–2499. [CrossRef]

29. Katis, I.N.; He, P.J.; Eason, R.W.; Sones, C.L. Improved sensitivity and limit-of-detection of lateral flow devices using spatial constrictions of the flow-path. *Biosens. Bioelectron.* **2018**, *113*, 95–100. [CrossRef]

30. Ross, G.; Bremer, M.G.; Wichers, J.H.; Van Amerongen, A.; Nielen, M.W. Rapid Antibody Selection Using Surface Plasmon Resonance for High-Speed and Sensitive Hazelnut Lateral Flow Prototypes. *Biosensors* **2018**, *8*, 130. [CrossRef]

31. Ekins, R.; Wild, D. Chapter 2.5—Ambient analyte assay. In *The Immunoassay Handbook*, 4th ed.; Elsevier: Oxford, UK, 2013; pp. 109–121.

32. Yetisen, A.K.; Akram, M.S.; Lowe, C.R. Paper-based microfluidic point-of-care diagnostic devices. *Lab Chip* **2013**, *13*, 2210–2251. [CrossRef] [PubMed]

33. Oh, Y.K.; Joung, H.-A.; Kim, S.; Kim, M.-G. Vertical flow immunoassay (VFA) biosensor for a rapid one-step immunoassay. *Lab Chip* **2013**, *13*, 768–772. [CrossRef] [PubMed]

34. Reuterswärd, P.; Gantelius, J.; Andersson Svahn, H. An 8 min colorimetric paper-based reverse phase vertical flow serum microarray for screening of hyper IgE syndrome. *Analyst* **2015**, *140*, 7327–7334. [CrossRef] [PubMed]

35. Rey, E.G.; O'Dell, D.; Mehta, S.; Erickson, D. Mitigating the Hook Effect in Lateral Flow Sandwich Immunoassays Using Real-Time Reaction Kinetics. *Anal. Chem.* **2017**, *89*, 5095–5100. [CrossRef]

36. Tate, J.; Ward, G. Interferences in immunoassay. *Clin. Biochem. Rev.* **2004**, *25*, 105–120.

37. Clarke, O.J.R.; Goodall, B.L.; Hui, H.P.; Vats, N.; Brosseau, C.L. Development of a SERS-Based Rapid Vertical Flow Assay for Point-of-Care Diagnostics. *Anal. Chem.* **2017**, *89*, 1405–1410. [CrossRef]

38. Eltzov, E.; Marks, R.S. Colorimetric stack pad immunoassay for bacterial identification. *Biosens. Bioelectron.* **2017**, *87*, 572–578. [CrossRef]

39. Samsonova, J.V.; Safronova, V.A.; Osipov, A.P. Rapid flow-through enzyme immunoassay of progesterone in whole cows' milk. *Anal. Biochem.* **2018**, *545*, 43–48. [CrossRef]

40. Chinnasamy, T.; Segerink, L.I.; Nystrand, M.; Gantelius, J.; Andersson Svahn, H. Point-of-Care Vertical Flow Allergen Microarray Assay: Proof of Concept. *Clin. Chem.* **2014**, *60*, 1209–1216. [CrossRef]

41. Burmistrova, N.A.; Rusanova, T.Y.; Yurasov, N.A.; Goryacheva, I.Y.; De Saeger, S. Multi-detection of mycotoxins by membrane based flow-through immunoassay. *Food Control* **2014**, *46*, 462–469. [CrossRef]

42. Joung, H.-A.; Ballard, Z.S.; Ma, A.; Tseng, D.K.; Teshome, H.; Burakowski, S.; Garner, O.B.; Di Carlo, D.; Ozcan, A. Paper-based multiplexed vertical flow assay for point-of-care testing. *Lab Chip* **2019**, *19*, 1027–1034. [CrossRef] [PubMed]

43. Reese, I.; Holzhauser, T.; Schnadt, S.; Dölle, S.; Kleine-Tebbe, J.; Raithel, M.; Worm, M.; Zuberbier, T.; Vieths, S. Allergen and allergy risk assessment, allergen management, and gaps in the European Food Information Regulation (FIR). *Allergo J. Int.* **2015**, *24*, 180–184. [CrossRef] [PubMed]

44. Bremer, M.G.E.G.; Smits, N.G.E.; Haasnoot, W. Biosensor immunoassay for traces of hazelnut protein in olive oil. *Anal. Bioanal. Chem.* **2009**, *395*, 119–126. [CrossRef] [PubMed]

45. Walker, M.J.; Burns, D.T.; Elliot, C.T.; Gowland, M.H.; Mills, C.E.N. Is food allergen analysis flawed? Health and supply chain risks and a proposed framework to address urgent analytical needs. *Analyst* **2016**, *141*, 24–35. [CrossRef]

46. Croote, D.; Quake, S.R. Food allergen detection by mass spectrometry: The role of systems biology. *NPJ Syst. Biol. Appl.* **2016**, *2*, 16022. [CrossRef]

47. Zhao, Y.; Wang, H.; Zhang, P.; Sun, C.; Wang, X.; Wang, X.; Yang, R.; Wang, C.; Zhou, L. Rapid multiplex detection of 10 foodborne pathogens with an up-converting phosphor technology-based 10-channel lateral flow assay. *Sci. Rep.* **2016**, *6*, 21342. [CrossRef]

Biosensors **2019**, *9*, 143

48. Anfossi, L.; Di Nardo, F.; Giovannoli, C.; Passini, C.; Baggiani, C. Increased sensitivity of lateral flow immunoassay for ochratoxin A through silver enhancement. *Anal. Bioanal. Chem.* **2013**, *405*, 9859–9867. [CrossRef]
49. Raeisossadati, M.J.; Danesh, N.M.; Borna, F.; Gholamzad, M.; Ramezani, M.; Abnous, K.; Taghdisi, S.M. Lateral flow based immunobiosensors for detection of food contaminants. *Biosens. Bioelectron.* **2016**, *86*, 235–246. [CrossRef]
50. Taylor, S.B.; Christensen, G.; Grinter, K.; Sherlock, R.; Warren, L. The Allergen Bureau VITAL Program. *J. AOAC Int.* **2018**, *101*, 1–5. [CrossRef]
51. Tsagkaris, A.S.; Nelis, J.L.D.; Ross, G.M.S.; Jafari, S.; Guercetti, J.; Kopper, K.; Zhao, Y.; Rafferty, K.; Salvador, J.P.; Migliorelli, D.; et al. Critical assessment of recent trends related to screening and confirmatory analytical methods for selected food contaminants and allergens. *TrAC Trends Anal. Chem.* **2019**, *121*, 115688. [CrossRef]

 *biosensors*

Article

Ex Vivo Raman Spectrochemical Analysis Using a Handheld Probe Demonstrates High Predictive Capability of Brain Tumour Status

Danielle Bury [1], Camilo L. M. Morais [1], Katherine M. Ashton [2], Timothy P. Dawson [2] and Francis L. Martin [1,*]

1   School of Pharmacy and Biomedical Sciences, University of Central Lancashire, Preston PR1 2HE, UK; deb11@doctors.org.uk (D.B.); CDLMedeiros-de-morai@uclan.ac.uk (C.L.M.M.)
2   Neuropathology, Royal Preston Hospital, Lancashire Teaching Hospitals NHS Trust, Sharoe Green Lane, Preston PR2 9HT, UK; Katherine.ashton@lthtr.nhs.uk (K.M.A.); timothy.dawson@lthtr.nhs.uk (T.P.D.)
*   Correspondence: flmartin@uclan.ac.uk

Received: 9 March 2019; Accepted: 29 March 2019; Published: 30 March 2019

Abstract: With brain tumour incidence increasing, there is an urgent need for better diagnostic tools. Intraoperatively, brain tumours are diagnosed using a smear preparation reported by a neuropathologist. These have many limitations, including the time taken for the specimen to reach the pathology department and for results to be communicated to the surgeon. There is also a need to assist with resection rates and identifying infiltrative tumour edges intraoperatively to improve clearance. We present a novel study using a handheld Raman probe in conjunction with gold nanoparticles, to detect primary and metastatic brain tumours from fresh brain tissue sent for intraoperative smear diagnosis. Fresh brain tissue samples sent for intraoperative smear diagnosis were tested using the handheld Raman probe after application of gold nanoparticles. Derived Raman spectra were inputted into forward feature extraction algorithms to build a predictive model for sensitivity and specificity of outcome. These results demonstrate an ability to detect primary from metastatic tumours (especially for normal and low grade lesions), in which accuracy, sensitivity and specificity were respectively equal to 98.6%, 94.4% and 99.5% for normal brain tissue; 96.1%, 92.2% and 97.0% for low grade glial tumours; 90.3%, 89.7% and 90.6% for high grade glial tumours; 94.8%, 63.9% and 97.1% for meningiomas; 95.4%, 79.2% and 98.8% for metastases; and 99.6%, 88.9% and 100% for lymphoma, based on smear samples ($\kappa = 0.87$). Similar results were observed when compared to the final formalin-fixed paraffin embedded tissue diagnosis ($\kappa = 0.85$). Overall, our results have demonstrated the ability of Raman spectroscopy to match results provided by intraoperative smear diagnosis and raise the possibility of use intraoperatively to aid surgeons by providing faster diagnosis. Moving this technology into theatre will allow it to develop further and thus reach its potential in the clinical arena.

Keywords: brain tumour diagnosis; classification; forward feature extraction algorithm; intraoperative use; Raman spectroscopy; Raman probe

1. Introduction

Brain tumours account for 3% of all tumours diagnosed annually [1]. Whilst this comprises a small proportion of total cancer burden, the difficulty of complete removal of the tumour is inherent. High-grade tumours can be infiltrative and when operating within the brain the risk of removing crucial structures in a bid to free the patient of the tumour, yet risk leaving them with significant neural deficit is ever present. Up to 75% of tumour resections are thought to leave behind viable tumour, though there is a survival benefit to improved/complete resection [2,3]. Therefore, any new technique available to highlight residual tumour, thus improving outcome and resection, yet reducing the non-tumour tissue removed would be beneficial. Currently, the use of 5-aminolevulinic acid (5-ALA) does allow for fluorescence of tumour cells in order to aid resection; however, this is imperfect. It can be difficult to tell apart tumour from background fluorescence [4].

In recent years many studies have been performed using vibrational spectroscopy in an effort to improve and decrease time to cancer diagnosis and aid resection of tumours. Vibrational spectroscopy includes two complementary techniques: Raman and attenuated total reflection Fourier-transform infrared (ATR-FTIR) spectroscopy. Raman spectroscopy detects chemical bonds via scattering of photons due to bond vibrations, whereas ATR-FTIR spectroscopy measures energy absorbance after excitation by an IR beam following reflection of the beam via an internal element (usually crystal) [5]. Both generate a 'fingerprint' of the elements within the examined sample, which can be examined to determine differences between them [5]. The majority of these studies have been ex vivo, with a move in recent years to increase the number of in situ studies [6], though these have yet to demonstrate definitive results. The need to test fresh tissue is crucial, to overcome any spectral changes seen due to formalin fixation or freezing artefact [7,8]. The use of gold nanoparticles in conjunction with Raman spectroscopy, known as surface enhanced Raman scattering (SERS) has previously been shown to improve the Raman signal received, reducing signal-to-noise ratio (SNR) and thus enhance the spectral quality [9]. This method uses molecules adsorbed onto the target surface prior to spectral acquisition, it has previously shown promise when used particularly with blood products to detect cancer [10].

The ability to use a probe intraoperatively, for example in brain surgery, and tell the surgeon in real-time if the tissue is cancerous or not would be greatly beneficial and perhaps the most useful area for Raman spectroscopy to make its clinical entry. Many areas within the cancer care pathway have been considered for targeting by spectroscopy, yet this is likely to be the best target location [10,11]. Stables et al. proposed a sound method to enable the surgeon to detect differences in the brain tissue found using spectroscopy as a method to provide real-time feedback [12]. This is an interesting suggestion, and certainly there is a need to develop technology to provide the surgeon with an answer without the need to interpret spectra. Desroches et al. used a handheld Raman probe intraoperatively with an accuracy of 87% to determine brain tumour from non-tumour tissue [13]. They then followed the study with the development of an optical biopsy needle for use during brain tumour biopsies. Following validation in an animal model, they tested their system during human brain surgery with an accuracy of 84%, sensitivity of 80% and specificity of 90% for tumour detection. These results were from comparing Raman spectra to biopsy results where the majority of the biopsy comprised tumour tissue [14]. This shows exciting potential; however, due to light contamination much of their study required procedures to be performed in darkness. Other types of Raman probes, such as based on stimulated Raman scattering, have been used in many biological applications, in particular for intracellular sensing and imaging [15].

Handheld probes have also been used for lymph node, breast and cervical testing. Horsnell et al. demonstrated that a handheld Raman probe used to determine the presence of cancer within sentinel lymph nodes with suspected breast cancer metastasis. They achieved sensitivities and specificities of up to 92% and 100% respectively, using frozen tissue [16]. They then went on to test lymph nodes using Raman micro-spectroscopy, and achieved concordance with histopathology in up to 91% of cases, improving as more points were assessed [17]. Within breast pathology, Haka et al. demonstrated a 93% accuracy in determining normal breast from benign or cancerous lesions [18]. They also demonstrated the potential of using Raman spectroscopy for intraoperative assessment of mastectomy margins with positive results, and possibly may have improved intraoperative results had spectroscopy been used in real-time [19]. As Raman spectroscopy is unaffected by aqueous materials therefore it is felt by the authors to be most suited to examining fresh brain tissue.

This study has been designed in order to determine the potential of the use of intraoperative SERS for brain tumour diagnosis within the neuropathology department. Raman spectral analysis of fresh brain tissue sent for intraoperative smear diagnosis was performed. The objective was to compare SERS results to both the intraoperative smear result and final formalin-fixed paraffin embedded tissue

(FFPE) result. This was done to understand if SERS can aid the clinical pathway and provide results similar to conventional neuropathology, with the aim of replacing the need for an intraoperative smear diagnosis, allowing the surgeon to test tissue intraoperatively to guide diagnosis and resection in the future.

2. Materials and Methods

Prior to using the handheld Raman machine, a custom-built box (sample compartment) was required to ensure darkness when analysing the tissues. As this was being placed into a working laboratory, it would not be possible to work in darkness and it would also need to fit into a category 2 fume hood for work with fresh tissue. With this is mind, a box was custom engineered using plywood. A stage was built within this box to allow the slide to be moved in the x and y planes with a custom cut out area for the slide to be held securely. This was to allow the tissue to be accurately positioned under the probe. A clamp was then secured to the box to allow the probe to be moved in the z plane to allow it to be positioned at the correct height above the tissue. Thus allowing movement similar to a conventional light microscope. The box was painted with black paint on the inside to minimise reflection of any light entering it. It also enabled it to be wiped clean if required. This was designed to be a prototype hence the materials involved (Figure 1).

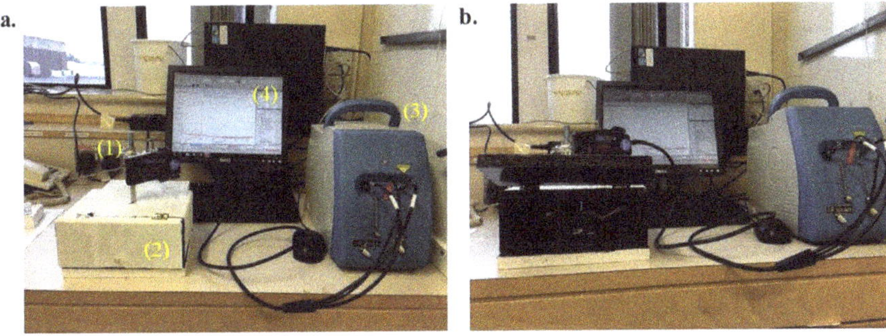

Figure 1. The handheld Raman probe in situ in the Neuropathology Department at Royal Preston Hospital. (**a**) Instrument setup showing (**1**) Raman probe, (**2**) sample compartment, (**3**) Raman detector and laser source, and (**4**) computer module with appropriate software. (**b**) A different view of the instrument setup with the sample compartment opened. (**c**) Sample holder with an example slide inside the sample compartment.

Integrating such studies in a typical clinical setting is a major challenge, especially during such serious and complicated procedures. Fresh brain tissue samples sent to the laboratory for intraoperative smear preparations were tested over a 6-month period (Table 1). Ethical approval was obtained from the BTNW brain bank (NRES14/EE/1270). We obtained $n = 29$ samples (a decent cohort size), which were analysed using an i-Raman portable Raman system with BAC100/BAC102 lab-grade Raman probe from B&W Tek from Pacer International, with software version 4.1. All consented samples arriving in the laboratory were tested over a six-month time period, typical of a routine clinical setting.

Table 1. Results of both intraoperative smear preparations and final formalin-fixed paraffin-embedded tissue for each case tested.

Case Number	Smear Result	Paraffin Result
1	Low-grade glioma	Glioblastoma
2	Meningioma	Meningioma
3	Metastasis	Ovarian serous carcinoma
4	High-grade glioma	Glioblastoma
5	High-grade glioma	Glioblastoma
6	Meningioma	Meningioma
7	Metastasis	Adenocarcinoma
8	High-grade glioma	Glioblastoma
9	High-grade glioma	Glioblastoma
10	Metastasis	Renal cell carcinoma
11	Metastasis	Lung adenocarcinoma
12	no tumour	Glioblastoma
13	Low-grade glioma	Astrocytoma Grade 2
14	Inflammation	Astrocytoma Grade 2
15	Inflammation	Astrocytoma Grade 2
16	Metastasis	Ovarian serous carcinoma
17	High-grade glioma	Glioblastoma
18	High-grade glioma	Glioblastoma
19	High-grade glioma	Glioblastoma
20	High-grade glioma	Glioblastoma
21	High-grade glioma	Glioblastoma
22	reactive Low-grade glioma	Low grade glioma
23	Intermediate-grade glioma	Glioblastoma
24	Low-grade glioma	Astrocytoma Grade 3
25	Lymphoma	High grade B cell lymphoma
26	Glioma	Astrocytoma Grade 2
27	No definite tumour	Astrocytoma Grade 2
28	Low- to intermediate-grade glioma	Astrocytoma Grade 2
29	High-grade glioma	Glioblastoma

The samples tested using the Raman spectrometer were obtained from tissue sent for intraoperative smear diagnosis. This tissue was then formalin fixed along with any remaining tissue for formal neuropathological examination. The sample arrived via air-tube from theatre within 5–10 min of removal from the patient. Prior to sample analysis, a small amount of tissue (similar in size to that used for a smear preparation) [20] was placed onto a glass slide covered with aluminium foil [21] and 100 μL of 5 μg/mL BioPureTM 20 nm gold nanoparticles diluted in PBS was dropped onto the sample and left for 2 min to absorb prior to collecting 10 spectra per sample. Gold nanoparticles were used to enhance spectral quality. Each spectrum had an acquisition time of 30 s at a laser power of 75%, field of view 0.9 mm × 0.9 mm, with a 785 nm laser. In total the spectra took 5 min to acquire and 2 min to save prior to analysis. This equates to the time taken to prepare a smear preparation prior to histopathological analysis. Once analysed tissue was formalin-fixed for final histopathological diagnosis.

Data analysis was then conducted using MATLAB R2014b software (MathWorks Inc., Natick, MA, USA) with an IRootlab toolkit [22]. The raw spectral data were initially pre-processed by cutting the region of interest, 1800–400 cm^{-1}, followed by polynomial baseline correction and vector normalisation. Thereafter, principal component analysis-linear discriminant classifier (PCA-LDC) was applied for classification of the datasets on a spectral basis. The training and validation sets were split using sub-dataset generation specification algorithm within IRootlab toolkit, where the model validation was performed using 10% of samples randomly assigned to the validation set during model construction. Due to the small number of samples, each spectrum was analysed separately for a predicted tissue pathology. PCA-LDC uses PCA as feature extraction method, where the original data is decomposed into a few number of principal components (PCs) representing the majority of the information in the original dataset. The scores on each PC are then used as input variables for linear discriminant analysis (LDA). LDA works by maximizing the between-class variance over the within-class variance in order to create a linear decision boundary between the classes that provides the optimum class segregation [23]. Patient factors, including the location of the tumour and biopsy were not considered within this study. Whilst the tumour site does influence the histopathological diagnosis, many pathologists prefer to start the diagnostic process on morphology only, blinded to demographics, history and site to prevent unconscious bias. Once a morphological diagnosis or range of differential diagnoses are formulated this can be tested against the site and demographics. Therefore, the Raman analysis is essentially being used to offer the same initial analysis as a pathologist on morphology only. Demographics, history and site are part of the secondary analysis, which as this study develops in the future could then be included within the analysis.

3. Results

Over the 29 samples from 27 patients, 290 spectra were collected and analysed. Due to the relatively small number of samples, each spectrum was analysed separately for a predicted tissue pathology. From this, PCA-LDC was employed and receiver operating characteristic (ROC) curves generated. This was done to determine the classification accuracies of the Raman spectra as compared to both the intraoperative smear result and final FFPE histological diagnosis, followed by ROC curves to determine the accuracy of the classification model as well as its sensitivity and specificity were generated. Low-grade gliomas were considered WHO grades 1 and 2, and high-grade gliomas WHO grades 3 and 4. Meningiomas were classed as WHO grade 1. Metastatic tumours were grouped due to the range of different primary sites within the tumours tested, and as intraoperatively 'metastasis' is sufficient for intraoperative surgical planning. The Raman spectra with or without gold nanoparticles for the same type of sample (high-grade glioma) are shown in Figure 2a. SERS is emitted from only the molecules adsorbed on the nanoparticles surface; thus the spectrum is not always the same as the corresponding Raman spectrum without nanoparticles. In our case, although the shape of both spectra looks similar, the Raman spectrum with gold nanoparticles contains higher intensities in the regions between ~1300–1700 cm^{-1} and ~700–1000 cm^{-1}. Figure 2b depicts the Raman spectra for non-tumour brain and cancer (high-grade glioma) tissue samples in presence of gold nanoparticles.

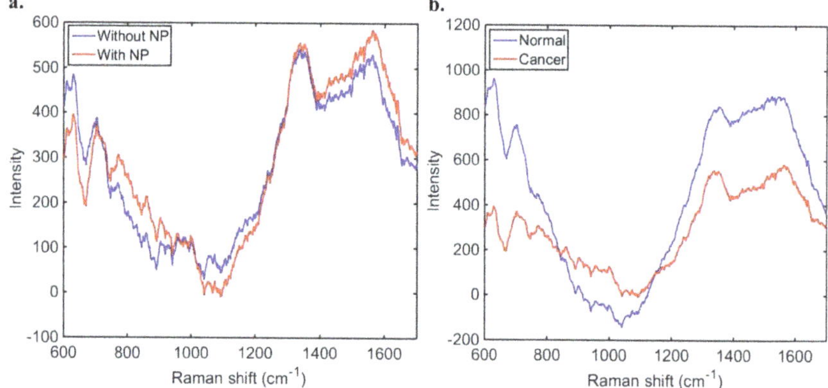

Figure 2. Baseline-corrected Raman spectra for (**a**) high-grade glioma tissues with and without gold nanoparticles (NP); (**b**) none tumour (normal) brain and cancer (high-grade glioma) tissues with gold nanoparticles. Each spectrum represents the average of 10 measurements in the tissue sample.

3.1. SERS Results Compared to Intraoperative Smear Preparation

From Figure 3 it can be seen that the accuracy for detection of primary brain tumours was between 64% and 92%. The algorithm provided the lowest accuracy for meningioma (64%) with differentiation of glial tumours proving more robust (92.2 and 89.7%). The ROC parameters and curves (Figure 4, Table 2) demonstrate the sensitivities and specificities range from 64%–94% and 91%–100%, respectively, again with meningioma falling behind the other tumours for sensitivity. As the area under the curve is >0.8 for all tumour classifications it confirms the high accuracy of the classification model and presence of statistical significance ($P < 0.001$). This is an important result if this model is to provide clinically useful information. With the exception of meningioma the positive and negative predictive values are consistently high (Table 2), with all negative predictive values over 95%.

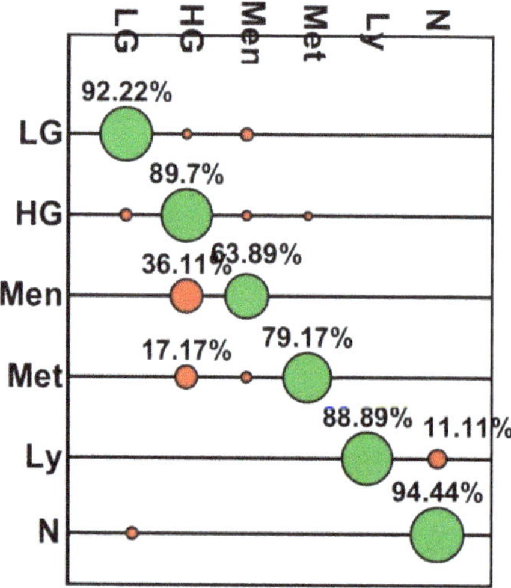

Figure 3. Graphical confusion matrix for PCA-LDC model using smear-based results. Key: N; Non tumour brain tissue, LG; Low-grade Glioma, HG; High-grade Glioma, Men; Meningioma, Met: Metastasis, Ly; Lymphoma. Green demonstrates those correctly classified, whereas red indicates an incorrect classification.

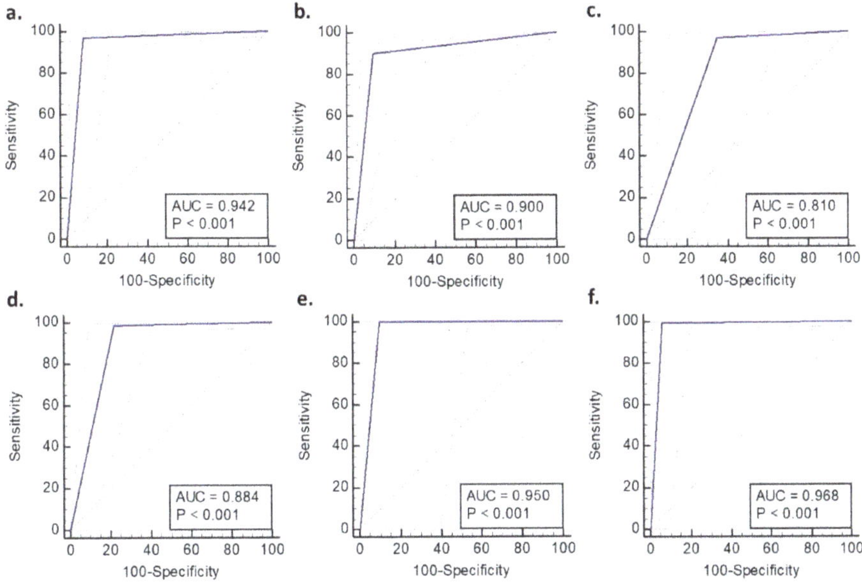

Figure 4. Receiver operating characteristic curves for smear-based samples: (**a**) Low-grade Glioma; (**b**) High-grade Glioma; (**c**) Meningioma; (**d**) Metastasis; (**e**) Lymphoma; and, (**f**) Non tumour brain tissue. Dashed pale blue curves represent 95% confidence intervals (AUC: area under the curve).

Table 2. Figures of merit for PCA-LDC model using smear-based samples. Cohen's kappa coefficient (κ) = 0.87.

Class	Accuracy (%)	Sensitivity (%)	Specificity (%)	PPV (%)	NPV (%)
N	98.6	94.4	99.5	97.7	98.8
LG	96.1	92.2	97.0	88.7	98.0
HG	90.3	89.7	90.6	83.5	94.4
Men	94.8	63.9	97.1	62.1	97.3
Met	95.4	79.2	98.8	93.3	95.8
Lv	99.6	88.9	100	100	99.6

Key: N; Non-tumour brain tissue, LG; Low-grade Glioma, HG; High-grade Glioma, Men; Meningioma, Met: Metastasis, Ly; Lymphoma, PPV; positive predictive value, NPV; negative predictive value.

3.2. SERS Results Compared to FFPE Tissue Results

When comparing the SERS results to the final FFPE diagnosis, the classification model also works with a high degree of accuracy. With the exception of metastatic tumours, the accuracy dips slightly for all cases as compared to the smear results (Figure 5, Table 3). This may be due to a variety of reasons, including non-tumour brain tissue within the biopsy material or areas of necrosis. Given this is not possible to determine macroscopically by eye, this remains a limitation of the study. The reduction in classification accuracy is to be expected as the neuropathologist has many diagnostic tools to aid the final FFPE diagnosis such as tumour morphology, architecture and immunohistochemical testing. The ROC graphs though do continue to show the reliability and statistical significance of the classification model (Figure 6), highlighting the ability of SERS to differentiate the tumour types within this study. Following formalin fixation all samples were found to contain tumour tissue therefore no non-tumour samples are represented within this analysis.

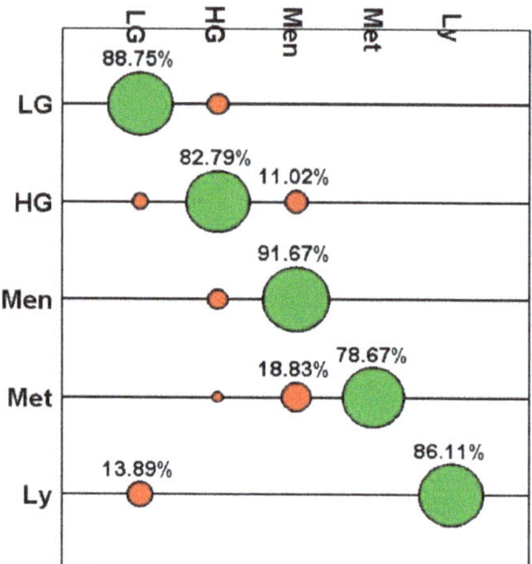

Figure 5. Graphical confusion matrix for PCA-LDC model using formalin fixed paraffin-embedded tissue results. Key: LG; Low-grade Glioma, HG; High-grade Glioma, Men; Meningioma, Met: Metastasis, Ly; Lymphoma. Green demonstrates those correctly classified, whereas red indicates an incorrect classification.

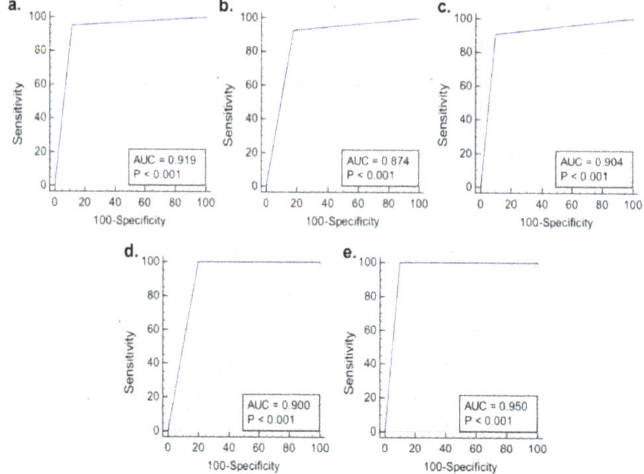

Figure 6. Receiver operating characteristic curves for formalin-fixed paraffin-embedded tissue results: (**a**) Low-grade Glioma; (**b**) High-grade Glioma; (**c**) Meningioma; (**d**) Metastasis; (**e**) Lymphoma. Dashed pale blue curves represent 95% confidence intervals (AUC: area under the curve).

Table 3. Figures of merit for PCA-LDC model using paraffin-embedded tissue results. Cohen's kappa coefficient (κ) = 0.85.

Class	Accuracy (%)	Sensitivity (%)	Specificity (%)	PPV (%)	NPV (%)
LG	93.8	88.7	95.4	85.8	96.4
HG	88.0	82.8	92.8	91.6	85.1
Men	90.8	91.7	90.8	42.4	99.3
Met	96.3	78.7	100	100	95.7
Lv	99.5	86.1	100	100	99.5

Key: LG; Low-grade Glioma, HG; High-grade Glioma, Men; Meningioma, Met: Metastasis, Ly; Lymphoma, PPV; positive predictive value, NPV; negative predictive value.

4. Discussion

Many Raman spectroscopic studies have been performed in recent years with the aim of introducing a clinically useful diagnostic tool that is easy to use and reagent-free. Much work has been performed towards standardisation of methodology and analysis, as this has previously led to criticism as many different techniques have been used [24]. Previous work within the field has shown good discrimination between normal and cancerous tissue. For example, within brain tumours, prostate and ovarian cancer we have previously found potential using Raman spectroscopy to differentiate normal from tumour within both tissue and biofluids [5,25–27]. The aim of this study was to determine if a handheld Raman probe could provide comparable results to both an intraoperative smear preparation and the final FFPE histological diagnosis. Comparable results would allow for further exploration of a Raman based probe for intraoperative use, particularly within the field of neuro-oncology. The use of fresh tissue, within a neuropathology laboratory, testing samples sent for smear preparations demonstrates a novel approach within this field, moving spectroscopic assessment closer to the patient. This study was designed to be a snap-shot of a typical clinical setting in a neuro-oncology surgical department over a 6-month period in order to ascertain the potential of extending our investigations into the potential translatability of this approach for routine practice.

These results demonstrate the ability of a handheld Raman device, when combined with gold nanoparticles, to differentiate tumour types from fresh brain tissue. The results are comparable to both the intraoperative smear preparations and final FFPE diagnosis, with accuracy at detecting a variety of primary brain tumours and metastases ranging from 63.9–94.4% as compared to the intraoperative smear preparation, and 78.7–91.7% when compared to the FFPE diagnosis. With the exception of meningioma the sensitivities and specificities are above 75% throughout, with the majority over 90%. The PPV and NPV results are also consistently high. It is possible that the meningioma group demonstrated lower accuracies due to varying morphological appearances. A much larger study would be required to determine differences between meningiomas of different types and grade. A similar issue applies to metastasis. This is a difficult group to combine as they are from different primary sites and therefore a different phenotype which will be expressed and would likely account for the differences in spectral classification. Due to the small number of metastatic tumours it was not possible to sub-classify these based on spectra within this study. This would be a next step when taking this study forward into a larger test group. These results are also comparable to a recent study demonstrating the possible use of Raman to detect tumours prior to biopsy [14]. For a test to be clinically useful, especially intraoperatively, a high accuracy, PPV and NPV is needed. These results compare well to a study performed on intraoperative smears and the final results compared to the FFPE diagnosis, which yielded an accuracy of 95.25% with PPV of 95.3% and NPV of 95.1% [28]. This is an important step as it allows the results to be comparative to current techniques, possibly demonstrating an improvement. By adequately training the Raman probe these results demonstrate a possible improvement on the current method of intraoperative smear diagnosis, reducing the human element involved and decreasing time to reach a diagnosis. As the accuracy of the Raman probe is slightly reduced when results are compared to the FFPE diagnosis for the majority of tumours (see Figures 3 and 5), the role for conventional neuropathology remains, with this tool focused towards intraoperative diagnosis. As this technique is taken forward, it would also require improvements within the Raman spectrometer to reduce the signal to noise ratio, in order for the nanoparticles to no longer be required, as one of the leading benefits to spectroscopy is the lack of labelling required. They were however felt to be important to this study given it is an early step in the introduction of spectroscopy into a surgical theatre.

These positive findings indicate the possible benefits to having a handheld Raman device present within the neurosurgical theatre, although much larger datasets need to be explored before clinical trial, in particular for classes that had a small number of samples in this study, such as lymphoma and metastasis. As all tissue was preserved following spectral acquisition and fixed to aid final diagnosis, we have also shown that spectral acquisition and addition of nanoparticles have not harmed the tissue, nor prevented final histological diagnosis as a final diagnosis by a Consultant Neuropathologist was possible in all cases. This is an important step when bringing this technology into the clinical field. Patient factors were not considered within this study as they were felt unlikely to directly influence the histopathological assessment. As the technique is developed, it may prove useful to add patient characteristics into an algorithm to improve accuracy, particularly within the paediatric field as some tumours are inherent to certain age groups.

When comparing these results to the final FFPE, it is understood that a final histopathological diagnosis includes many factors, including molecular analysis. The comparison was performed within this study to demonstrate differences primarily between an intraoperative result and the FFPE. Further studies would be required to determine if the Raman spectral differences were able to differentiate the underlying molecular changes, thus circumventing the need for molecular testing.

These results would suggest a handheld device within theatres, may be able to assist surgeons in removing tumour tissue without the need for an intraoperative smear preparation. This could reduce surgical time as no result is awaited and allow for improved surgical resection as small foci of tumour could be identified. It can be seen that the time taken to prepare and take the spectra is similar to that required to produce a smear preparation, therefore the time saved is within the analysis and removal of the need to send the sample to the pathology laboratory. Moving this study forward, it would be important to study the junction between non-tumour brain tissue, brain tissue infiltrated by tumour cells and tumour tissue to understand the threshold at which the Raman probe is able to detect tumour cells. This would therefore allow demonstration of any benefit of its use over current methods.

As the classification model is able to determine tumour type this also would allow for further management steps to be completed, such as the addition of Gliadel wafers in the case of high-grade gliomas. The use of intracranial chemotherapy, such as Gliadel, is recommended by the National Institute for Clinical Excellence (NICE) under certain conditions, one of which is the diagnosis intraoperatively of a high-grade glioma by a neuropathologist [29]. Raman spectroscopy could therefore be used to circumvent the need to involve the neuropathologist, streamlining processes within theatre. The identification of a metastatic tumour is also important when planning the level of resection undertaken. We have not used the results to determine primary tumour origin for metastatic tumours, as this has previously been shown to be challenging, particularly for cases such as adenocarcinomas from different primary sites [30]. Intraoperatively, the determination of a metastasis versus a primary brain tumour is the level required and offered from an intraoperative smear preparation. Therefore allowing conventional histopathology and immunohistochemistry to determine the primary site of origin is the most logical step.

Determination of surgical margins within breast cancer has been demonstrating using Raman spectroscopy [19]. If developed, our classification model may also allow for other surgical sites to determine presence of absence of tumour intraoperatively, again removing the need for intraoperative frozen sections to be performed and improve resection clearance. Additional information such as samples descriptive statistics (e.g., age, gender and tumour location) would benefit the interpretation of the analysis results to provide a more patient-driven diagnosis. Herein, these factors were not considered, which limits this paper to the specific features of the cohort analysed. Nevertheless, Raman spectroscopy combined with chemometric algorithms has shown great potential for tumour differentiation, evidenced on the high accuracies, sensitivities and specificities achieved for this data set.

5. Conclusions

Overall, this study presents a novel approach to intraoperative brain tumour diagnosis and is one of the first studies to report results on intraoperative fresh brain tumour samples. The next step is to move this technology into theatre and continue to develop the classification model to allow for real-time feedback to the surgeon and allow Raman technology to reach its full potential.

Author Contributions: Conceptualization, F.L.M., T.P.D. and D.B.; methodology, F.L.M., D.B., K.M.A.; software, C.L.M.M.; formal analysis, C.L.M.M. and D.B.; investigation, K.M.A.; writing—original draft preparation, D.B. and

C.L.M.M.; writing—review and editing, F.L.M. and T.P.D.; supervision, F.L.M. and T.P.D; project administration and principal investigator, F.L.M.; funding acquisition, F.L.M. and T.P.D.

Funding: This research was funded by Rosemere Cancer Foundation, grant number RCF0313TPD. CLMM would like to thank CAPES-Brazil (grant 88881.128982/2016-01) for financial support.

Acknowledgments: The authors would like to acknowledge support from the Brain Tumour North West RTB and the Sidney Driscoll Neuroscience Foundation. We would also like to acknowledge the help and skill of Mr C Almond in the development and engineering of the custom-built box.

Conflicts of Interest: The authors declare no conflict of interest. The funders had no role in the design of the study; in the collection, analyses, or interpretation of data; in the writing of the manuscript, or in the decision to publish the results.

References

1. Brain, Other CNS and Intracranial Tumours Statistics. Available online: http://www.cancerresearchuk.org/health-professional/cancer-statistics/statistics-by-cancer-type/brain-other-cns-and-intracranial-tumours#heading-Zero (accessed on 18 February 2018).
2. Hollon, T.; Lewis, S.; Freudiger, C.W.; Xie, S.; Orringer, D.A. Improving the accuracy of brain tumour surgery via Raman-based technology. *Neurosurg. Focus* **2016**, *40*, E9–E25. [CrossRef]
3. Broadbent, B.; Tseng, J.; Kast, R.; Noh, T.; Brusatori, M.; Kalkanis, S.N.; Auner, G.W. Shining light on neurosurgery diagnostics using Raman spectroscopy. *J. Neurooncol.* **2016**, *130*, 1–9. [CrossRef] [PubMed]
4. Galli, R.; Uckermann, O.; Temme, A.; Leipnitz, E.; Meinhardt, M.; Koch, E.; Schackert, G.; Steiner, G.; Kirsch, M. Assessing the efficiency of coherent anti-Stokes Raman scattering microscopy for the detection of infiltrating glioblastoma in fresh brain tissue. *J. Biophotonics* **2017**, *10*, 404–414. [CrossRef] [PubMed]
5. Owens, G.L.; Gajjar, K.; Trevisan, J.; Fogarty, S.W.; Taylor, S.E.; Da Gama-Rose, B.; Martin-Hirsch, P.L.; Martin, F.L. Vibrational biospectroscopy coupled with multivariate analysis extracts potentially diagnostic features in blood plasma/serum of ovarian cancer patients. *J. Biophotonics* **2014**, *7*, 200–209. [CrossRef] [PubMed]
6. UK Hospital to trial Raman Probe for Brain Tumours. Available online: www.optics.org/news/6/1/18 (accessed on 22 February 2018).
7. O'Faolain, E.; Hunter, M.; Byrne, J.; Kellehan, P.; McNamara, M.; Byrne, H.; Lyng, F. A Study Examining the Effects of Tissue Processing on Human Tissue Sections using Vibrational Spectroscopy. *Vib. Spectrosc.* **2005**, *38*, 121–127. [CrossRef]
8. Huang, Z.; McWilliams, A.; Lam, S.; English, J.; McLean, D.I.; Lui, H.; Zeng, H. Effect of formalin fixation on the near-infrared Raman Spectroscopy of normal and cancerous human bronchial tissues. *Int. J. Oncol.* **2003**, *23*, 649–655. [CrossRef]
9. Butler, H.J.; Fogarty, S.W.; Kerns, J.G.; Martin-Hirsch, P.L.; Fullwood, N.J.; Martin, F.L. Gold nanoparticles as a substrate in bio-analytical near-infrared surface-enhanced Raman spectroscopy. *Analyst* **2015**, *140*, 3090–3097. [CrossRef] [PubMed]
10. Velicka, M.; Pucetaite, M.; Urboniene, V.; Ceponkus, J.; Jankevicius, F.; Sablinskas, V. Detection of cancerous kidney tissue by means of SERS spectroscopy of extracellular fluid. *J. Raman Spectrosc.* **2017**, *48*, 1744–1754. [CrossRef]
11. Bury, D.; Martin-Hirsch, P.L.; Martin, F.L.; Dawson, T.P. Are new technologies translatable to point-of-care testing? *Lancet* **2017**, *390*, 2765–2766. [CrossRef]
12. Stables, R.; Clemens, G.; Butler, H.J.; Ashton, K.M.; Brodbelt, A.; Dawson, T.P.; Fullwood, L.M.; Jenkinson, M.D.; Baker, M.J. Feature driven classification of Raman spectra for real time spectral brain tumour diagnosis using sound. *Analyst* **2016**, *142*, 98–109. [CrossRef] [PubMed]
13. Desroches, J.; Jerymn, M.; Mok, K.; Lemieux-Leduc, C.; Mercier, J.; St-Arnaud, K.; Urmey, K.; Guiot, M.-C.; Marple, E.; Petrecca, K.; et al. Characterization of a Raman spectroscopy probe system for intraoperative brain tissue classification. *Biomed. Opt. Exp.* **2015**, *6*, 2380–2397. [CrossRef] [PubMed]
14. Desroches, J.; Jerymn, M.; Pinto, M.; Picot, F.; Tremblay, M.-A.; Obaid, S.; Urmey, K.; Trudel, D.; Soulez, G.; Guiot, M.-C.; et al. A new method using Raman spectroscopy for in vivo targeted brain cancer tissue biopsy. *Sci. Rep.* **2018**, *8*, 1792–1802. [CrossRef] [PubMed]
15. Li, Y.; Wang, Z.; Mu, X.; Ma, A.; Guo, S. Raman tags: Novel optical probes for intracellular sensing and imaging. *Biotechnol. Adv.* **2017**, *35*, 168–177. [CrossRef]

16. Horsnell, J.; Stonelake, P.; Christie-Brown, J.; Shetty, G.; Hutchings, J.; Kendall, C.; Stone, N. Raman spectroscopy—A new method for the intra-operative assessment of axillary lymph nodes. *Analyst* **2010**, *135*, 3042–3047. [CrossRef]
17. Horsnell, J.D.; Smith, J.A.; Sattlecker, M.; Sammon, A.; Chrisite-Brown, J.; Kendall, C.; Stone, N. Raman spectroscopy—A potential new method for the intra-operative assessment of axillary lymph nodes. *Surgeon* **2012**, *10*, 123–127. [CrossRef]
18. Haka, A.S.; Volynskaya, Z.; Gardecki, J.A.; Nezemi, J.; Shenk, R.; Wang, N.; Dasari, R.R.; Fitzmaurice, M.; Feld, M.S. Diagnosing breast cancer using Raman spectroscopy: prospective analysis. *J. Biomed. Opt.* **2009**, *14*, 054023. [CrossRef]
19. Haka, A.S.; Volynskaya, Z.; Gardecki, J.A.; Nazemi, J.; Lyons, J.; Hicks, D.; Fitzmaurice, M.; Dasari, R.R.; Crowe, J.P.; Feld, M.S. In vivo Margin Assessment during Partial Mastectomy Breast Surgery Using Raman Spectroscopy. *Cancer Res.* **2006**, *66*, 3317–3322. [CrossRef]
20. Ellison, D.; Love, S.; Chimelli, L.; Harding, B.N.; Lowe, J.S.; Vinters, H.V.; Brandner, S.; Yong, W.H. *Neuropathology: A reference text of CNS Pathology*, 3rd ed.; Elsevier Mosby: St. Louis, MO, USA, 2013.
21. Cui, L.; Butler, H.J.; Martin-Hirsch, P.L.; Martin, F.L. Aluminium foil as a potential substrate for ATR-FTIR, transflection FTIR or Raman spectrochemical analysis of biological samples. *Anal. Meth.* **2016**, *8*, 481–487. [CrossRef]
22. Trevisan, J.; Angelov, P.P.; Scott, A.D.; Carmichael, P.L.; Martin, F.L. IRootLab: A free and open-source MATLAB toolbox for vibrational biospectroscopy data analysis. *Bioinformatics* **2013**, *29*, 1095–1097. [CrossRef] [PubMed]
23. Santos, M.C.D.; Morais, C.L.M.; Nascimento, Y.M.; Araujo, J.M.G.; Lima, K.M.G. Spectroscopy with computational analysis in virological studies: A decafe (2006–2016). *Trends Anal. Chem.* **2017**, *97*, 244–256. [CrossRef]
24. Butler, H.J.; Ashton, L.; Bird, B.; Cinque, G.; Curtis, K.; Dorney, J.; Esmonde-White, K.; Fullwood, N.J.; Gardner, B.; Martin-Hirsch, P.L.; et al. Using Raman spectroscopy to characterize biological material. *Nat. Protoc.* **2016**, *11*, 664–687. [CrossRef] [PubMed]
25. Gajjar, K.; Heppenstall, L.D.; Pang, W.; Ashton, K.M.; Trevisan, J.; Patel, I.I.; Llabjani, V.; Stringfellow, H.F.; Martin-Hirsch, P.L.; Dawson, T.; et al. Diagnostic segregation of human brain tumours using Fourier-transform infrared and/or Raman spectroscopy coupled with discriminant analysis. *Anal. Meth.* **2012**, *5*, 89–102. [CrossRef] [PubMed]
26. Patel, I.I.; Trevisan, J.; Singh, P.B.; Nicholson, C.M.; Gopala Krishnan, R.K.; Matanhelia, S.S.; Martin, F.L. Segregation of human prostate tissues classified high-risk (UK) versus low risk (India) for adenocarcinoma using Fourier-transform infrared or Raman microspectroscopy coupled with discriminant analysis. *Anal. Bioanal. Chem.* **2011**, *401*, 969–982. [CrossRef]
27. Bury, D.; Morais, C.L.M.; Paraskevaidi, M.; Ashton, K.M.; Dawson, T.P.; Martin, F.L. Spectral classification for diagnosis involving numerous pathologies in a complex clinical setting: A neuro-oncology example. *Spectrochim. Acta A Mol. Biomol. Spectrosc.* **2019**, *206*, 89–96. [CrossRef] [PubMed]
28. Sanjeev, K.; Aparna, B.; Anuradha, K.; Brijesh, T.; Sanjay, K.; Neetika, S. Intraoperative Squash Cytology of Central Nervous System and Spinal Cord Lesions with Histological Correlation. *Arch. Pathol Lab. Med.* **2016**, *3*, 61–72.
29. Carmustine Implants and Temozolomide for the Treatment of Newly Diagnosed High-Grade Glioma. Technology appraisal guidance. Available online: https://www.nice.org.uk/Guidance/TA121 (accessed on 21 February 2018).
30. Krafft, C.; Shapoval, L.; Sobottka, S.B.; Geiger, K.D.; Schackert, G.; Salzer, R. Identification of primary brain metastasis by SIMCA classification of IR spectroscopic images. *Biochim. Biophys. Acta* **2006**, *1758*, 883–891. [CrossRef]

Article

Robust SERS Platforms Based on Annealed Gold Nanostructures Formed on Ultrafine Glass Substrates for Various (Bio)Applications

Lan Zhou [1], Simone Poggesi [1,2], Giuliocesare Casari Bariani [1,2], Rakesh Mittapalli [1], Pierre-Michel Adam [1], Marisa Manzano [2] and Rodica Elena Ionescu [1,*]

[1] Light, Nanomaterials and Nanotechnology (L2N), FRE-CNRS 2019, Institute Charles Delaunay (ICD), University of Technology of Troyes, 12 Rue Marie Curie CS 42060, 10004 Troyes CEDEX, France; lan.zhou@utt.fr (L.Z.); poggesi.simone@spes.uniud.it (S.P.); casaribariani.giuliocesare@spes.uniud.it (G.C.B.); rakesh.mittapalli@utt.fr (R.M.); pierre_michel.adam@utt.fr (P.-M.A.)

[2] Dipartimento di Scienze Agroalimentari, Ambientali e Animali (DI4A), Università degli Studi di Udine, Via Sondrio 2/A, 33100 Udine, Italy; marisa.manzano@uniud.it

* Correspondance: elena_rodica.ionescu@utt.fr; Tel.: +33-3-2575-9728; Fax: +33-3-2571-8456

Received: 14 February 2019; Accepted: 29 March 2019; Published: 10 April 2019

Abstract: In this study, stable gold nanoparticles (AuNPs) are fabricated for the first time on commercial ultrafine glass coverslips coated with gold thin layers (2 nm, 4 nm, 6 nm, and 8 nm) at 25 °C and annealed at high temperatures (350 °C, 450 °C, and 550 °C) on a hot plate for different periods of time. Such gold nanostructured coverslips were systematically tested via surface enhanced Raman spectroscopy (SERS) to identify their spectral performances in the presence of different concentrations of a model molecule, namely 1,2-bis-(4-pyridyl)-ethene (BPE). By using these SERS platforms, it is possible to detect BPE traces (10^{-12} M) in aqueous solutions in 120 s. The stability of SERS spectra over five weeks of thiol-DNA probe (2 µL) deposited on gold nano-structured coverslip is also reported.

Keywords: SERS on ultrafine solid supports; glass coverslips; BPE; thiol-DNA probe; annealed gold nanostructures

---

1. Introduction

In recent decades, the use of gold nanoparticles (AuNPs) in the field of light–matter interactions has attracted considerable interest for their potential applications in various sciences, such as biomedical, agricultural, environmental, and forensic investigations, because of their unique optical and chemical properties. AuNPs serve as miniaturized platforms, ideal for the development of ultrasensitive bioassays [1,2]. In fact, a considerable number of protocols have been developed for the preparation of AuNPs, which can be classified into three main groups: (i) the top-down approach based on physical manipulation, for example using an ultrasonic field, electron beam lithography [3], or laser irradiation [4]; (ii) the bottom-up method based on the chemical reduction of chloroauric acid to AuNPs in the presence of reducing and stabilizing agents [5], and (iii) the "on solid supports" approach, using an annealed microscope glass slide coated with thin gold film [6–10].

Among the optical methods, surface enhanced Raman spectroscopy (SERS) has attracted scientists' attention, being used in the identification of unknown substances in analytical chemistry [11], electrochemistry [12], physical chemistry [13], solid state physics, biochemistry, biophysics, and even medicine [14–16]. Nowadays, SERS effects on metal (Au)-coated surfaces are explained using electromagnetic and chemical mechanisms [17,18]. The SERS electromagnetic mechanism is caused by the interactions between the laser excitation on the metal-labeled surface and the scattered Raman field. The chemical mechanism is caused by the inelastic tunneling of ballistic electrons to the lowest unoccupied molecular orbital of the chemisorbed molecule. The return of the electron to its initial state in the metal—i.e., the recombination of the electron and the hole—emits a Raman-shifted photon [19–21]. Usually, SERS substrates are fabricated either by immobilizing a colloidal silver nanoparticle (AgNP) on 3-aminopropyltriethoxysilane-coated glass coverslips [22] or by dropping tiny volumes of colloidal AgNPs onto microscope glass coverslips [23–25]. Despite the simplicity, such SERS substrates are not stable, and AgNPs are easily displayed by water streams. On the contrary, for biological applications, naked AgNPs must be strongly attached to transparent and biocompatible solid supports for further use in different chemical and biomolecule functionalization steps without any nanoparticle displacements. To solve the inconvenience of the stability of nanoparticles on solid supports, a solution was reported in 2013, which consisted of heating the gold-coated microscope glass pieces at a high annealing temperature in an oven for 8 h [6]. However, these substrates require training to carefully cut the microscope slide into small pieces to avoid scratching that may affect the homogeneity in the AgNP formation.

A second solution is proposed in the present work and consists of replacing microscope glass pieces with ultrafine coverslips, thus eliminating the cutting step. It should be noted that glass coverslips are typically used for running conventional biological assays and have never been used for robust SERS (bio)applications. The aim of this work is therefore to validate the use of ultrafine glass coverslips as easy-to-handle and inexpensive SERS supports after a high annealing treatment on a hot plate for several hours. A wide variety of chemicals and biomolecules can be detected with these new SERS platforms. To prove the concept, a Raman model molecule, 1,2-bis-(4-pyridyl)-ethene (BPE) [26,27], was selected to study the SERS spectroscopic performances of annealed gold nanostructures on ultrafine glass coverslips.

2. Materials and Methods

2.1. Materials

The coverslips were cleaned using Decon 90 (Decon Laboratories™ Decon 90™) liquid detergent (Fisher Scientific, Göteborg, Sweden) and ultrapure water (18.2 MΩ cm) produced by a Millipore Milli-Q water purification system (Molsheim, France). The same water was used for all rinsing steps.

For SERS investigations, a BPE (1,2-bis-(4-pyridyl)-ethene) molecule was purchased from Sigma-Aldrich (Schnelldorf, Germany). Several BPE solutions were prepared from 97% concentrated stock solution, to form six concentrations that were subsequently tested in ultrapure water: 10^{-3}, 10^{-5}, 10^{-7}, 10^{-9}, 10^{-12}, and 10^{-15} M, respectively. For SERS stability studies, a fragment of DNA modified in 5′ position with C6 thiol group (TGTTTGAGCGTCATTTCCTTCTCACTATTTAGTGGTTATGAGATTACACGAGG, 53 pb), provided by Eurofins Genomics (Eberseberg, Germany) and here called thiol-DNA probe (10 ng/µL), was suspended in 1xSSPE buffer containing 3 M sodium chloride, 0.23 M sodium phosphate dibasic, 25 mM ethylenediaminetetraacetic acid, pH 7.4. The thiol-DNA was designed to detect *Brettanomyces bruxellensis* spoilage yeast. All the reagents required for the preparation of the SSPE buffer were provided by Sigma-Aldrich.

2.2. Instruments for the Characterization of Gold Nanoparticles Annealed on Coverslips

Metal evaporation was performed with Plassys MEB 400 (Plassys, Bestek, France). A hot plate (Thermo Fisher Scientific, Waltham, MA, USA) was used for annealing under clean room conditions.

Nanostructured coverslips were characterized with a scanning electron microscope (SEM) (FEG-SU8030, Tokyo, Japan) and an atomic force microscope (AFM) (Bruker ICON, Billeric, MA, USA) with cantilever ScanAsyst-Air in silicon nitride with a tip height of 2.5–8.0 mm. A spring constant of 4 N/m and a reflective aluminum coating on the back side in standard ScanAsyst-Air mode were used to characterize the morphology of AuNPs (data not shown).

SERS spectra were recorded with backscattering geometry using a modified Jobin-Yvon LabRAM (Horiba scientific, Longjumeau, France) and an excitation wavelength of 632.8 nm (11 mW) from the He–Ne laser source, and all the spectra were recorded with a 10× objective Olympus MPlanFl with a 5.2 μm^2 laser spot area. The acquisition time varied from 10 to 120 s, and all the spectra were recorded 3 times with a D filter range between 0 and 0.3.

For sterilization, a Tuttnauer Autoclave Steam Sterilizer 2540ML (Tuttnauer, Villenoy France) was used. The samples were dried in an oven provided by VWR company (DRY-Line drying oven DL 53), and all operations were made under a biological hood provided by Thermo-scientific MSC 1,2 ADV (Illkirch Cedex, France).

2.3. Sample Preparation: Cleaning, Gold Evaporation, and Annealing of Coverslips

Glass coverslips (Carl Roth GmbH + Co., KG, Karlsruhe, Germany) were degreased with Millipore distilled water and a detergent solution (Decon 90) (ratio 2:8, *v/v*) in an ultrasonic distilled water bath (Elmasonic S30H model, Elma Schmidbauer GmbH, Singen, Germany) at 50 °C for 15 min according to the procedure used by Jia et al. [6]. In addition, an ultrasonic bath was made with distilled water at 50 °C for 5 min. The next step was to carefully rinse each coverslip with distilled water, dry them under a stream of nitrogen, and deposit them on a hot plate at 100 °C for 10 min. Further, the coverslips were labelled with a scotch band on an external side for correct handling, fixed on a circular evaporation plate (200 mm diameter), and finally exposed to gold vapors in the evaporator. Different gold thicknesses (2 nm, 4 nm, 6 nm, and 8 nm, respectively) were evaporated on squared glass coverslips at 1×10^{-5} Torr pressure at 25 °C using an evaporation rate of 0.03 nm/s. The resulting gold-coated glasses (4 sets of 12 coverslips/set) were systematically heated on a hot plate preheated to three different temperatures (350 °C, 450 °C, and 550 °C) for different time periods (1, 3, 6, and 9 h, respectively) (Figure 1).

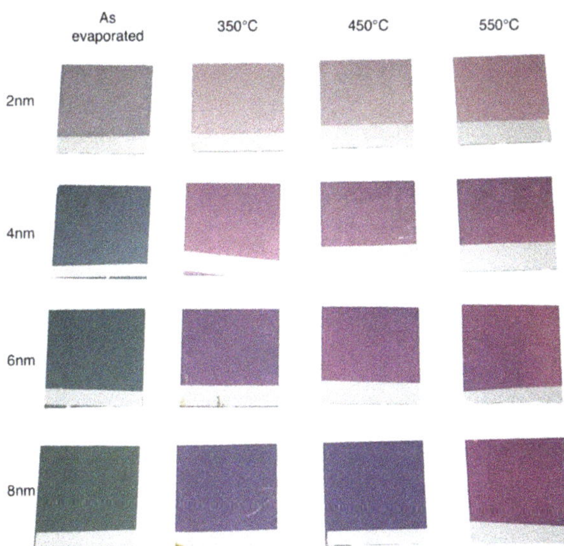

Figure 1. Square glass coverslips coated with gold thin films (2 nm, 4 nm, 6 nm, and 8 nm) after 3 h at three different temperatures (350 °C, 450 °C, and 550 °C).

After the annealing procedure, the coverslips underwent an additional cleaning process according to the procedure describe by Jia et al. [6], which involves washing with 70% ethanol in an ultrasonic bath at 30 °C for 20 min, rinsing with sterile water, and further washing in an ultrasonic bath with sterile water for 10 min at 30 °C. Then, the coverslips were allowed to dry in oven at 50 °C for 20 min.

After that, the annealed glass coverslips were biofunctionalized by adding 10 µL of thiol-DNA at 10 ng/µL in 1xSSPE at 4 °C overnight. The coverslips were subsequently washed with 1.5 mL of sterile water and dried over the biohood. The thiol-DNA was previously treated with a buffer solution containing 10 mM Tris(2-CarboxyEthyl)Phosphine hydrochloride) (TCEP) and 3 M sodium acetate in order to release the thiol group.

2.4. SERS Measurements on Coverslips

Different BPE concentrations were tested (10^{-3}, 10^{-5}, 10^{-7}, 10^{-9}, and 10^{-12} M) by deposing tiny drops of 2 µL on gold coverslips. In order to increase the sensitivity of the SERS experiments, different combinations of spectral acquisition time and laser filtering were used: for 10^{-3} M and 10^{-5} M, the acquisition time was 10 s using the D0 filter, whereas for the lower concentrations, a D0.3 filter was used to avoid the background noise due to the longer excitation time necessary for comparable spectra acquisition. An acquisition time of 30–120 s was studied. The stability of SERS spectra over five weeks for gold nanostrutured coverslips modified with thiol-DNA probe with an acquisition time of 10–30 s and using a D0.3 filter is also reported.

3. Results and Discussion

3.1. SEM Characterization

It is well known that the SERS properties of gold nanostructures are strongly influenced by the size, distribution, and spacing between particles [8]. In the preparation of the AuNP process, including cleaning, evaporation, and annealing protocol, the morphology of the substrate can be modulated by controlling different experimental conditions, such as the Millipore and Decon 90 distilled water ratio, gold film thickness, evaporation pressure, evaporation rate, annealing time, and annealing temperature. In our experiments, three parameters—the thickness of the evaporated gold film, the annealing temperature, and the annealing time—played an important role in the morphology of the gold nanoparticles and SERS properties. Thus, SEM studies were conducted for four different gold film thicknesses (2 nm, 4 nm, 6 nm, and 8 nm, respectively) at three different annealing temperatures (350 °C, 450 °C, and 550 °C) and for four different annealing times (1, 3, 6, and 9 h) (Figure 2).

3.1.1. Influence of the Annealing Temperature on the Formation of Gold Nanoparticles

Evaporated gold films of 2 nm, 4 nm, 6 nm, and 8 nm on coverslips showed different colors, from light blue (2 nm Au) to blue (4 nm Au), to light green (6 nm Au), or to darker green (8 nm Au). These colors changed significantly for each gold thickness after 3 h of annealing at different temperatures. The highest temperature produced a violet color for the 2 nm gold film, whereas for the 4 nm, 6 nm, and 8 nm films, the color appeared from light violet to dark purple, respectively (Figure 1). SEM images of the evaporated samples and the annealed samples are shown in Figure 2. The size of the gold nanoparticles increased with the increase of the thickness of the film (2 nm, 4 nm, 6 nm, and 8 nm), which corresponded to the color variation before and after annealing at different temperatures.

For glass coverslips coated with 2 nm Au, the interparticle distances, or proportion of background (PB), increased when temperatures rose from 350 °C to 550 °C (Figure 3). On the contrary, for samples coated with 6 nm and 8 nm Au, the PB values at 550 °C ranged from 60.94% to 63.63%, while at 450 °C, the PB values were 61.39% and 65.53%, respectively. Interestingly, the PB values for the 4 nm Au sample showed no great variation when annealed at 350 °C (60.29%), 450 °C (60.07%), or 550 °C (60.41%).

In conclusion, the temperature definitively influenced the sizes and shapes of the gold nanoparticles and the interparticle distances, with respect to the gold thickness evaporated on the coverslips.

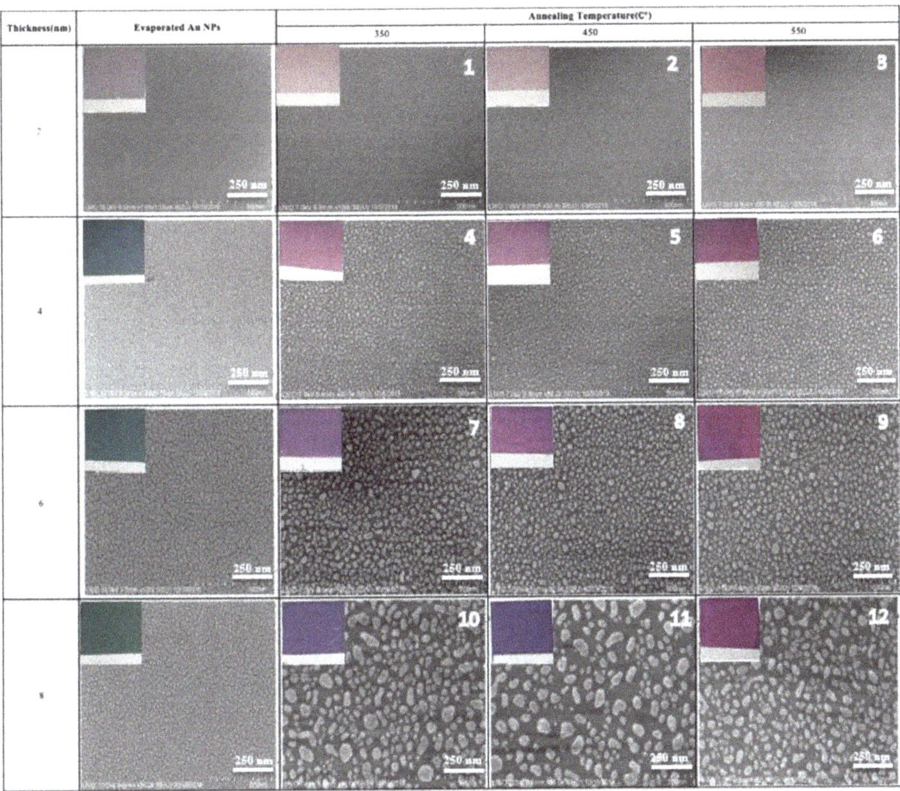

Figure 2. SEM images of square glass coverslips gold coated (2 nm, 4 nm, 6 nm, and 8 nm) after 3 h at different temperatures (350 °C, 450 °C, and 550 °C). AuNPs: gold nanoparticles.

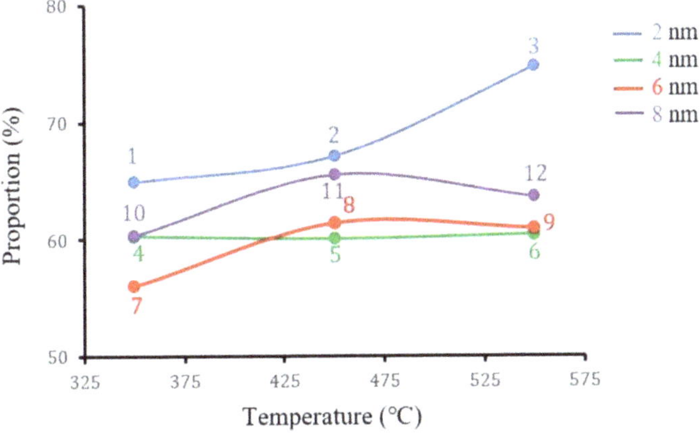

Figure 3. The proportion of background for the annealed square glass coverslips gold coated (2 nm, 4 nm, 6 nm, and 8 nm, respectively) after exposure at three different temperatures (350 °C, 450 °C, and 550 °C). These numbers are also used in Figure 2 to indicate the SEM image for every substrate.

3.1.2. Influence of Gold Thickness on Coverslips on Nanoparticle Distribution

As robust and stable SERS platforms, square coverslips coated with gold of 2 nm, 4 nm, 6 nm, and 8 nm were proposed and annealed at 550 °C for 3 h. For these samples, the particle size distribution, and the proportion of background are reported in Figure 4A,B.

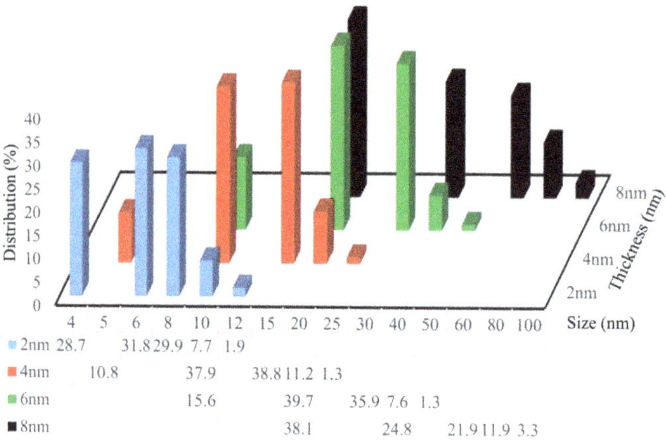

Size (nm)	4	5	6	8	10	12	15	20	25	30	40	50	60	80	100
2nm	28.7		31.8	29.9	7.7	1.9									
4nm		10.8			37.9		38.8	11.2	1.3						
6nm					15.6			39.7		35.9	7.6	1.3			
8nm								38.1			24.8		21.9	11.9	3.3

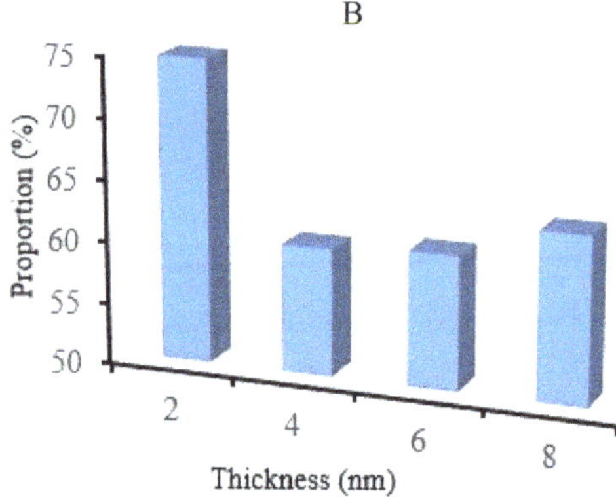

Figure 4. The size distribution of AuNPs on coverslips (**A**) and the proportion of background for different gold thicknesses (2 nm, 4 nm, 6 nm, and 8 nm) after annealing at 550 °C for 3 h (**B**).

Figure 4A shows that by increasing the thickness of gold, the size of the AuNP nanoparticles and their distribution percentage increase. Similarly, the size of nanoparticles affects the interdistance between particles. Thus, after the annealing of 2 nm Au film on the glass, the AuNPs ranged mainly from 6 (size distribution 31.8%) to 8 nm (29.9%), whereas for 4 nm Au, the nanoparticles ranged from 10 (37.9%) to 15 nm (38.8%). For 6 nm Au, the AuNPs ranged from 20 (39.7%) to 30 nm (35.9%), and finally, for 8 nm Au, the AuNPs ranged from 20 (38.1%) to 40 nm (24.8%).

On the other hand, it was found that the proportion of background for the coverslips coated with four different gold thicknesses (Figure 4B) was the smallest for 4 nm Au (60.41%) and the highest for 2 nm Au (74.77%). Additionally, for coverslips coated with 6 nm and 8 nm Au, the background was 60.94% and 63.63%, respectively.

In the current SERS studies, the 4 nm Au coated coverslips after 550 °C showed the largest nanoparticle surface coverage and the lowest interparticle distances compared with the other tested thicknesses (2 nm, 6 nm, and 8 nm, respectively).

3.1.3. Influence of Annealing Time on the Nanostructuration of Coverslips

In order to prepare large-scale gold nanoparticles with stable optical characteristics, the effect of 4 nm Au evaporated and annealed at 550 °C at different annealing times (1, 3, 6, and 9 h) is illustrated in Figure 5. On other hand, the nanoparticle size distribution and the proportion of background for SEM images (Figure 2) are analyzed using the public domain ImageJ software platform, developed at National Institutes of Health (Figure 6).

Experimentally, the coverslips heated for 1 h at 550 °C formed nanoparticles in the range of 15–20 nm (36.8–23%). Similar sample evolution was obtained for glasses after 6 h at the same temperature when the AuNPs ranged from 10 (23.6%) to 15 nm (38.8%), while the AuNP size after 3 h displayed a uniform distribution from 10 (37.9%) to 15 nm (38.8%) compared with the others, corresponding to the SEM image (Figure 5).

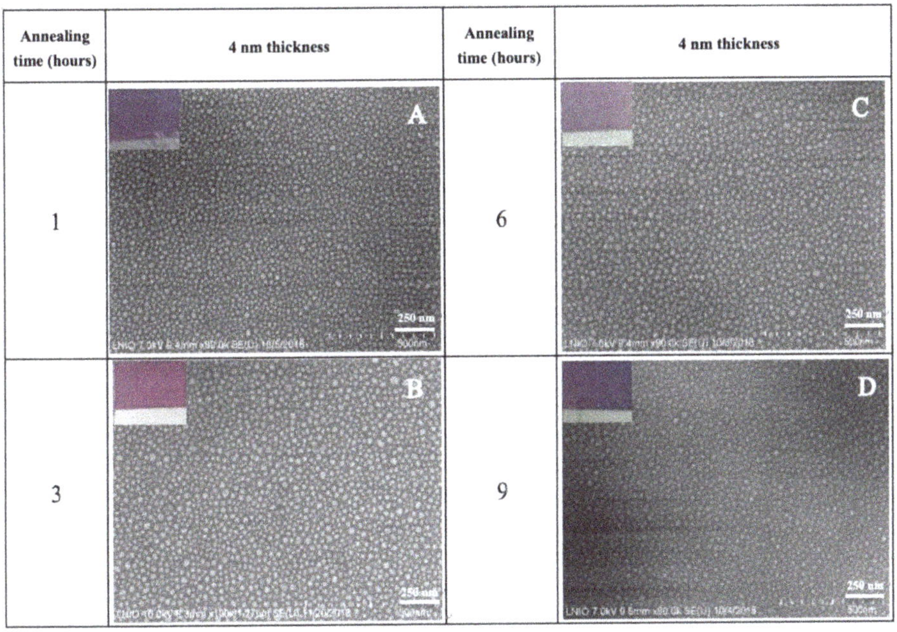

Annealing time (hours)	4 nm thickness	Annealing time (hours)	4 nm thickness
1	A	6	C
3	B	9	D

Figure 5. SEM images of AuNPs on square glass coverslips coated with 4 nm and annealed for different time periods (**A**) 1 h, (**B**) 3 h, (**C**) 6 h, and (**D**) 9 h at 550 °C.

The coverslips annealed for 9 h showed a high distribution at 5–10 nm (45.8–19.0%) (see Figure 6A). However, as shown in Figure 6B, the proportion of background increased following the evaporated gold film thickness, becoming thicker over time: 9 h (70.72%) > 6 h (64.87%) > 3 h (61.19%) >1 h (56.51%). In detail, even though the sample annealed for 9 h had a very high distribution of 45.8% at 5 nm, the largest proportion of the background of the sample was 70.72%, which corresponded to the coverage of the smallest area of the 9 h sample. On the other hand, the samples annealed for 1 h and 3 h had lower proportions compared with the samples annealed for 6 h and 9 h. The lower proportion of background of the larger surface coverage was obtained, and considering the uniform distribution of the nanoparticles and better surface coverage, we used the samples annealed at 550 °C for 3 h.

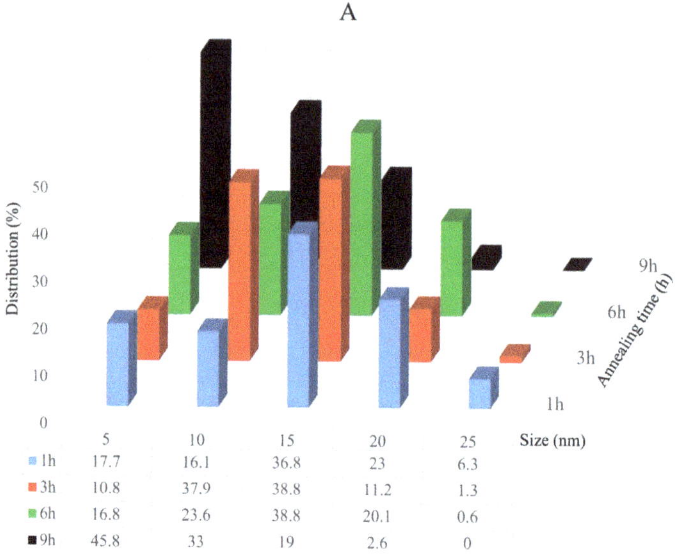

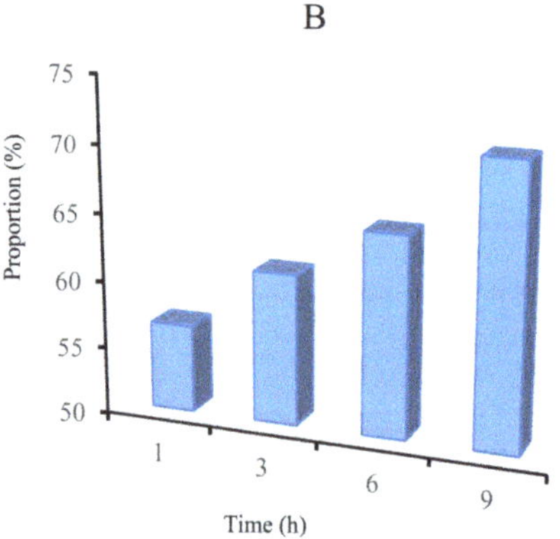

Figure 6. Analysis of AuNPs based on SEM images showing the size distribution of gold nanoparticles on annealed coverslips (**A**) and the proportion of background after annealing the 4 nm gold coated coverslips for different time periods (1, 3, 6, and 9 h) at 550 °C (**B**).

3.2. SERS Characterization

The SERS tests were initially performed on classical microscope glass slide supports modified with 4 nm gold film and annealed at 550 °C in the oven for 8 h according to the procedure described by Jia et al. [6]. These substrates proved to be inappropriate for SERS measurements, because the glass slide produced strong fluorescence interferences that abnormally altered the optical signals. Therefore, several solid supports are here proposed for SERS investigations: plastic petri dishes, glass coverslips, plastic pipettes, Eppendorf tubes, plastic cuvettes, and quartz crystals microbalance (QCM) (Supplementary Materials, Figure S1). The SERS measurements show that the best solid supports were the ultrafine glass coverslips for further gold nanostructuration due to the absence of fluorescence interferences. Then, the SEM morphology also confirmed the evolution of SERS signals for annealed gold coated coverslips at 350 °C, 450 °C, and 550 °C for 3 h (Figure S2A–C). Among the different spectroscopic investigations (Figure S3), it was found that the best SERS substrate is the ultrafine square glass coverslip coated with 4 nm Au (Figure S3B) annealed at 550 °C for 3 h (Figure S2C), due to the absence of background SERS peaks that could mask the presence of specific SERS peaks produced by (bio)molecules once immobilized on nanoparticles.

3.2.1. SERS Spectrum of BPE Molecule on 4 nm Gold-Annealed Coverslip

The SERS tests were carried out in the presence of a model molecule, 1,2-bis-(4-pyridyl)-ethene, which has interesting bonds and atoms giving good SERS spectra when deposited on annealed gold coated coverslips, as demonstrated in Figure 7.

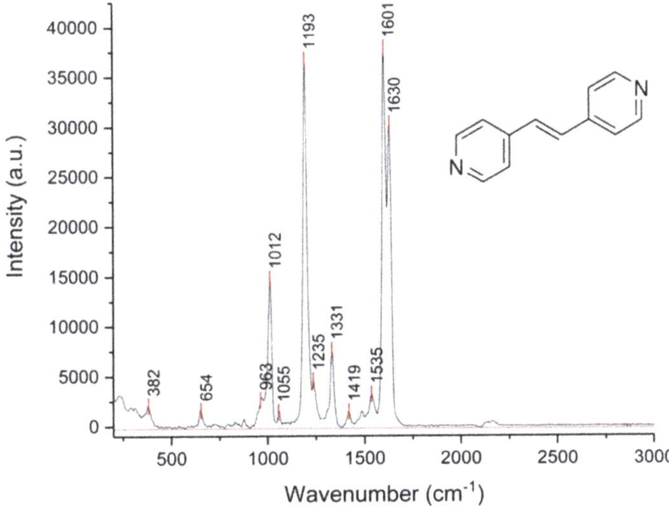

Figure 7. Surface enhanced Raman spectroscopy (SERS) spectrum of the 1,2-bis-(4-pyridyl)-ethene (BPE) (1 mM) on annealed gold nanostructured coverslip (4 nm Au, 550 °C for 3 h on a hot plate), after three times of acquisition of 10 s and using a D0 filter.

As reported, the main peaks at 1601 and 1630 cm^{-1} correspond to the C–N stretching mode in the pyridyl ring and the BPE vinyl group vibration, respectively [9], while the peaks at 1193 and 1235 cm^{-1} refer to the ring breathing mode of pyridine and the vibrational movement of the nitrogen atom in pyridyl, respectively. In the present work, the peak at 1012 cm^{-1} can be attributed to the chemical absorption of BPE molecules onto AuNPs on coverslips.

3.2.2. SERS Spectra of Different BPE Concentrations

SERS signals were recorded and compared for different BPE concentrations deposited on gold-annealed coverslips. These SERS measurements confirmed that the best conditions for the detection of very low BPE concentrations are as follows: 4 nm Au on glass heated at 550 °C for 3 h. As is shown in Figure 8, gold nanostructured coverslips made possible the SERS detection of the lower concentration of the BPE molecule at 10^{-12} M.

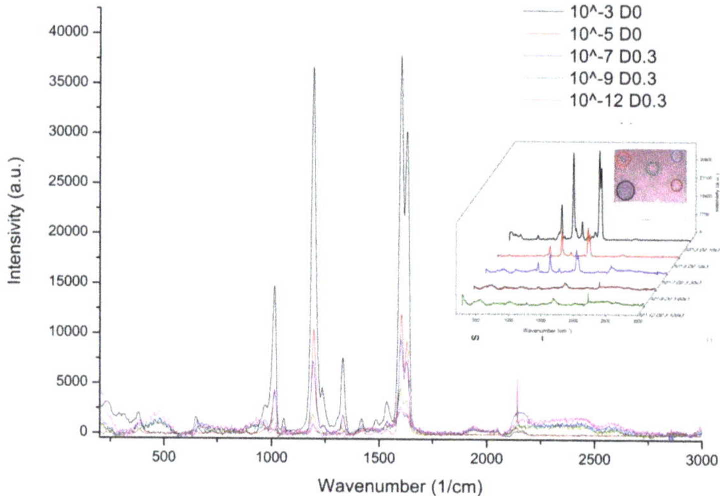

Figure 8. SERS spectra of BPE molecules of different concentrations (10^{-3}, 10^{-5}, 10^{-7}, 10^{-9}, and 10^{-12} M) using 4 nm gold-coated coverslips annealed at 550 °C for 3 h on a hot plate. Inset-photo of a coverslip after the deposition of five different BPE concentrations.

Figure 9 depicts the intensity variation of the three main SERS peaks (1193 cm^{-1}, 1630 cm^{-1}, and 1601 cm^{-1}) versus the decimal logarithmic of four BPE concentrations. Whatever the wavenumber, the intensity variation exhibited an autonomous decay when the concentration of BPE decreased. A linear fit was used to model the experimental SERS measurements. A pronounced linear behavior was observed for the 1630 wavenumber, for which a linear regression coefficient of 0.9976 was calculated. The values of the modeled slopes that represent the sensitivity were of the same order of magnitude.

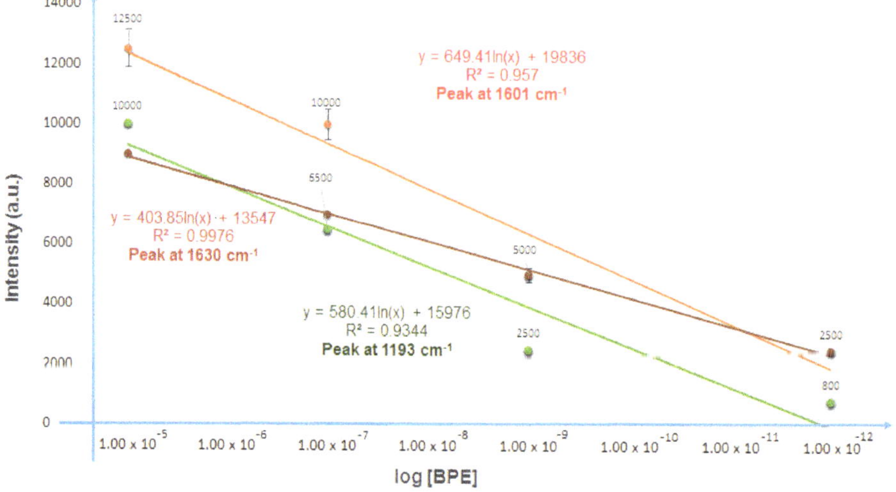

Figure 9. SERS intensity of three wavenumber main peaks (1193 cm^{-1}, 1630 cm^{-1}, and 1601 cm^{-1}) as a function of BPE concentration.

The enhancement factor (EF) was calculated using the equation EF = (I_{SERS}/I_R) × (N_R/N_{SERS}) and was found to be equal to 2.71×10^7. I_{SERS} represents the intensity of the 1630 cm^{-1} BPE band, measured for 10^{-5} M concentration, while I_R represents the intensity of the Raman band for 10^{-3} M on reference glass. N_{SERS} and N_R represent the number of molecules formed as a layer of 10 nm thickness under the laser spot and the number of the BPE molecules in the focal volume. The values of I_R and N_R are the same as those reported in [28]. Gold nanoparticles covered 40% of the surface under the spot. By using the same calculation method, N_{SERS} was found to be equal to 125 molecules for a surface spot laser of 5.2 μm^2. In the case of 10^{-5} M BPE content, an I_{SERS} value of 9000 was measured.

3.2.3. SERS Signal Stability of Thiol-DNA Deposited on Gold-Annealed Coverslip Substrate

A thiol-DNA biofunctionalized gold-annealed coverslip was tested over five weeks to evaluate the substrate's SERS signal stability (Figure 10). Interestingly, the intensity of SERS increased after two weeks and decreased after four weeks. This confirms that the nanostructuration of the coverslip was stable for more than a month. The stability tests were stopped after five weeks, because the thiol-DNA probe on AuNPs presented a strong attenuation of the SERS intensity.

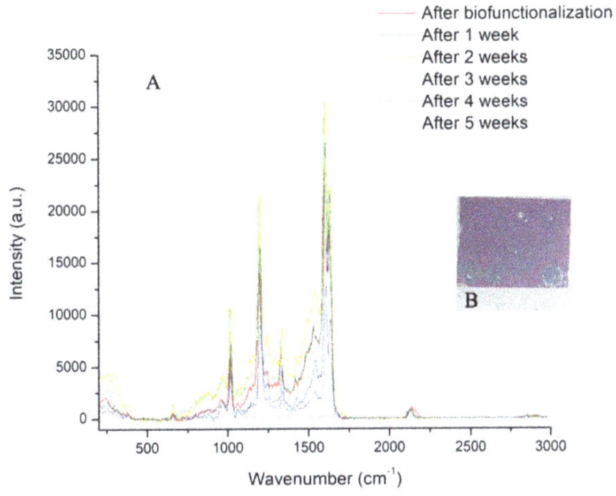

Figure 10. Evolution of SERS intensity of the thiol-DNA probe on annealed gold-coated coverslip over five weeks and obtained after three acquisition times, showing a 10 s spectra with a D0.3 filter (**A**). Photo of the sample after five weeks (**B**).

4. Conclusions

Large-scale, annealed, gold nanostructures were fabricated for the first time on ultrafine glass coverslips. Several parameters have been optimized to conclude that 4 nm gold-coated coverslip heated at 550 °C on a hot plate for 3 h had the greater sensitivity of the SERS spectrum to different BPE concentrations. By using the newly SERS annealed coverslips platforms it was possible to detect a BPE concentration of 10^{-12} M. Moreover, the stability of SERS spectra intensity over five weeks of a thiol-DNA probe (10 ng/μL) was also monitored. On the basis of these results, annealed gold coverslips can be considered as ideal substrates in the construction of ultrasensitive SERS nanobiosensors.

Supplementary Materials: The following are available online at http://www.mdpi.com/2079-6374/9/2/53/s1, Figure S1: SERS spectra of various naked solid supports: plastic petri dish, glass coverslip, plastic pipette, Eppendorf tube, plastic cuvette and quartz QCM crystal, Figure S2: SERS signals of naked annealed gold films (2, 4, 6 and 8 nm) on coverslips after 3 h, Figure S3: SERS signals of naked annealed gold films on coverslips at 550 °C for different periods.

Author Contributions: Conceptualization and methodology, R.E.I.; software, S.P., L.Z. and G.C.B.; validation, R.E.I., L.Z., and S.P.; formal analysis, R.E.I. and P.-M.A.; investigation, R.M.; resources and data curation, R.E.I. and L.Z.; writing—original draft preparation, R.E.I.; writing—review and editing, L.Z., M.M. and R.E.I.; visualization, P.-M.A.; supervision, project administration and funding acquisition, R.E.I.

Funding: This research was funded by EIPHI Graduate School grant number ANR17-EURE-0002.

Acknowledgments: R.E.I., G.C.B. and R.M. are thankful for the financier support of EIPHI Graduate School (contract "ANR17-EURE-0002"). The authors thank the NANOMAT Champagne-Ardenne Regional Platform for their financial support. L.Z. kindly thanks the Chinese Scholarship Council (CSC) for funding her PhD scholarship in France (October 2017–April 2021). S.P. thanks the Master thesis research program at the University of Udine for the financial support in France (September–December 2018).

Conflicts of Interest: The authors declare no conflicts of interest.

References

1. Saha, K.; Agasti, S.S.; Kim, C.; Li, X.N.; Rotello, V.M. Gold nanoparticles in chemical and biological sensing. *Chem. Rev.* **2012**, *112*, 2739–2779. [CrossRef] [PubMed]
2. Hu, J.Q.; Wang, Z.P.; Li, J.H. Gold nanoparticles with special shapes: Controlled synthesis, surface-enhanced Raman scattering, and the application in bio-detection. *Sensors* **2007**, *7*, 3299–3311. [CrossRef] [PubMed]
3. Galarreta, B.C.; Norton, P.R.; Lagugné-Labarthet, F. SERS detection of streptavidin/biotin monolayer assemblies. *Langmuir* **2011**, *27*, 1494–1498. [CrossRef]
4. Ratautas, K.; Gedvilas, M.; Voisiat, B.; Raciukaitis, G.; Grigonis, A. Transformation of a thin gold film to nanoparticles after nanosecond-laser irradiation. *J. Laser Micro Nanoeng.* **2012**, *7*, 355–361. [CrossRef]
5. Sarkar, J.; Roy, S.K.; Laskar, R.A.; Chattopadhyay, D.; Acharya, K. Bioreduction of chloaurate ions to gold nanoparticles by culture of *Pleurotcus sapidus* Quel. *Mater. Lett.* **2013**, *92*, 313–316. [CrossRef]
6. Jia, K.; Bijeon, J.L.; Adam, P.M.; Ionescu, R.E. Large scale fabrication of gold nano-Structured substrates via high temperature annealing and their direct use for the LSPR etection of atrazine. *Plasmonics* **2013**, *8*, 143–151. [CrossRef]
7. Jia, K.; Adam, P.M.; Marks, R.S.; Ionescu, R.E. Fixed Escherichia coli bacterial templates enable the production of sensitive SERS-based gold nanostructures. *Sens. Actuators B Chem.* **2015**, *211*, 213–219. [CrossRef]
8. Vaskvich, A.; Rubinstein, I. Localized Surface Plasmon Resonance (LSPR) transducers based on random evaporated gold island films: Properties and sensing applications. In *Nanoplasmonic Sensors*; Springer: New York, NY, USA, 2012.
9. Bonyár, A.; Csarnovics, I.; Veres, M.; Himics, L.; Csik, A.; Kámán, J.; Balázs, L.; Kökényesi, S. Investigation of the performance of thermally generated gold nanoislands for LSPR and SERS applications. *Sens. Actuators B Chem.* **2018**, *255*, 433–439. [CrossRef]
10. Serrano, A.; de la Fuente, O.R.; García, M.A. Extended and localized surface plasmons in annealed Au films on glass substrates. *J. Appl. Phys.* **2009**, *108*, 074303. [CrossRef]
11. Jiang, Y.; Sun, D.-W.; Pu, H.; Wei, Q. Ultrasensitive analysis of kanamycin residue in milk by SERS-based aptasensor. *Talanta* **2019**, *197*, 151–158. [CrossRef] [PubMed]
12. Sur, U.K.; Chowdhury, J. Surface-enhanced Raman scattering: Overview of a versatile technique used in electrochemistry and nanoscience. *Curr. Sci.* **2013**, *105*, 923–939.
13. Schlucker, S. Surface-enhanced Raman spectroscopy: Concepts and chemical applications. *Angew. Chem. Int. Ed.* **2014**, *53*, 4756–4795. [CrossRef]
14. Fikiet, M.A.; Khandasammy, S.R.; Mistek, E.; Ahmed, Y.; Halámková, L.; Bueno, J.; Lednev, I.K. Surface enhanced Raman spectroscopy: A review of recent applications in forensic science. *Spectrochim. Acta A* **2018**, *197*, 255–260. [CrossRef]
15. Szlag, V.M.; Rodriguez, S.; He, J.; Hudson-Smith, N.; Kang, H.; Le, N.; Reineke, T.M.; Haynes, C.L. Molecular affinity agents for intrinsic surface-enhanced Raman scattering (SERS) sensors. *ACS Appl. Mater. Interfaces* **2018**, *10*, 31825–31844. [CrossRef]
16. Ilkhani, H.; Hughes, T.; Li, J.; Zhong, C.J.; Hepel, M. Nanostructured SERS-electrochemical biosensors for testing of anticancer drug interactions with DNA. *Biosens. Bioelectron.* **2016**, *80*, 257–264. [CrossRef]
17. Yamamoto, Y.S.; Itoh, T. Why and how do the shapes of surface enhanced Raman scattering spectra change? Recent progress from mechanistic studies. *J. Raman Spectrosc.* **2016**, *47*, 78–88. [CrossRef]
18. Wang, A.X.; Kong, X. Review of recent progress of plasmonic materials and nano-structures for surface-enhanced Raman scattering. *Materials* **2015**, *8*, 3024–3052. [CrossRef]
19. Ren, B.; Liu, G.K.; Lian, X.B.; Yang, Z.L.; Tian, Z.Q. Raman spectroscopy on transition metals. *Anal. Bioanal. Chem.* **2007**, *388*, 29–45. [CrossRef]

20. Campion, A.; Kambhampati, P. Surface-Enhanced Raman Scattering. *Chem. Soc. Rev.* **1998**, *27*, 241–250. [CrossRef]
21. Kneipp, K. Surface-enhanced Raman scattering. *Phys. Today* **2007**, *60*, 40–46. [CrossRef]
22. Tharion, J.; Satija, J.; Mukherji, S. Facile synthesis of size-tunable silver nanoparticles by heteroepitaxial growth method for efficient NIR SERS. *Plasmonics* **2015**, *10*, 753–763. [CrossRef]
23. Zhai, Z.; Nie, M.; Guan, Y.; Zhang, F.; Chen, L.; Muhammad, D.W.; Liu, G.; Tian, Y.; Huang, Q. A microfluidic surface-enhanced Raman spectroscopy approach for assessing the particle number effect of AgNPs on cytotoxicity. *Ecotoxicol. Environ. Saf.* **2018**, *162*, 529–535. [CrossRef]
24. Culhane, K.; Jiang, K.; Neumann, A.; Pinchuk, A.O. Laser-fabricated plasmonic nanostructures for surface-enhanced Raman spectroscopy of bacteria quorum sensing molecules. *MRS Adv.* **2017**, *2*, 2287–2294. [CrossRef]
25. Suchomel, P.; Prucek, R.; Cerna, K.; Fargasova, A.; Panacek, A.; Gedanken, A.; Zboril, R.; Kvitek, L. Highly efficient silver particle layers on glass substrate synthesized by the sonochemical method for surface enhanced Raman spectroscopy purposes. *Ultrason. Sonochem.* **2016**, *32*, 165–172. [CrossRef]
26. Ou, F.S.; Hu, M.; Naumov, L.; Kim, A.; Wu, W.; Bratkovsky, A.M.; Li, X.; Williams, R.S.; Li, Z. Hot-spot engineering in polygonal nanofinger assemblies for surface enhanced Raman spectroscopy. *Nano Lett.* **2011**, *11*, 2538–2542. [CrossRef]
27. Herrera, G.M.; Padilla, A.C.; Hernandez-Rivera, S. Surface enhanced Raman scaterring (SERS) studies of gold and silver nanopartices prepared by laser ablation. *Nanomaterials* **2013**, *3*, 158–172. [CrossRef]
28. Khanafer, M.; Izquierdo-Lorenzo, I.; Akil, S.; Louarn, G.; Toufaily, J.; Hamich, T.; Adam, P.M; Jradi, S. Silver nanoparticles rings of controllable size multi-wavelenght SERS response and high enhancement of three pyridine derivatives. *ChemistrySelect* **2016**, *6*, 1201–1206. [CrossRef]

 *biosensors*

Article

The Application of a Nanomaterial Optical Fiber Biosensor Assay for Identification of *Brucella* Nomenspecies

Kelly McCutcheon [1], Aloka B. Bandara [2], Ziwei Zuo [1], James R. Heflin [1] and Thomas J. Inzana [2,3,*]

[1] Department of Physics, College of Science, Virginia Tech, Blacksburg, VA 24061, USA; kmccutch@vt.edu (K.M.); dzwei.dzwo@gmail.com (Z.Z.); rheflin@vt.edu (J.R.H.)
[2] Department of Biomedical Sciences and Pathobiology, Virginia-Maryland College of Veterinary Medicine, Virginia Tech, Blacksburg, VA 24061, USA; abandara@vt.edu
[3] Long Island University, Brookville, NY 11548, USA
* Correspondence: Thomas.Inzana@liu.edu; Tel.: +1-516-299-3685; Fax: +1-516-299-3867

Received: 22 March 2019; Accepted: 14 May 2019; Published: 21 May 2019

Abstract: Bacteria in the genus *Brucella* are the cause of brucellosis in humans and many domestic and wild animals. A rapid and culture-free detection assay to detect *Brucella* in clinical samples would be highly valuable. Nanomaterial optical fiber biosensors (NOFS) are capable of recognizing DNA hybridization events or other analyte interactions with high specificity and sensitivity. Therefore, a NOFS assay was developed to detect *Brucella* DNA from cultures and in tissue samples from infected mice. An ionic self-assembled multilayer (ISAM) film was coupled to a long-period grating optical fiber, and a nucleotide probe complementary to the *Brucella* IS711 region and modified with biotin was bound to the ISAM by covalent conjugation. When the ISAM/probe duplex was exposed to lysate containing ≥100 killed cells of *Brucella*, or liver or spleen tissue extracts from *Brucella*-infected mice, substantial attenuation of light transmission occurred, whereas exposure of the complexed fiber to non-*Brucella* gram-negative bacteria or control tissue samples resulted in negligible attenuation of light transmission. Oligonucleotide probes specific for *B. abortus*, *B. melitensis*, and *B. suis* could also be used to detect and differentiate these three nomenspecies. In summary, the NOFS biosensor assay detected three nomenspecies of *Brucella* without the use of polymerase chain reaction within 30 min and could specifically detect low numbers of this bacterium in clinical samples.

Keywords: *Brucella abortus*; *Brucella melitensis*; *Brucella suis*; optical fiber; biosensor; nucleotide probe; light transmission; diagnosis

---

1. Introduction

Brucellae are bacterial pathogens responsible for brucellosis of domestic and wild animals and are zoonotic pathogens for humans. *Brucella* spp. are small gram-negative, nonmotile, aerobic, and slow-growing coccobacilli. Despite the recognition of brucellae as a single genospecies based on DNA-DNA hybridization studies, they are systematically classified based on host specificity. The main terrestrial nomenspecies are *B. abortus* (cattle), *B. melitensis* (goats and sheep), *B. suis* (pigs), *B. canis* (dogs), *B. ovis* (sheep), and *B. neotomae* (woodrats) [1,2]. In addition, *Brucella* spp. can also be isolated from marine mammals [3,4]. Human infections are acquired by consuming unpasteurized milk and dairy products or by direct exposure to animals and their carcasses. Human brucellosis resulting from direct exposure is primarily a disease of farmers, shepherds, veterinarians, microbiologists, butchers, and slaughterhouse workers [5,6].

Wild animals play an important role in the epidemiology of *Brucella* infections. *Brucella* spp. remain enzootic in wild elk and bison in the Greater Yellowstone region that includes areas of Montana,

Idaho, and Wyoming. As a result, these animals are a reservoir for *B. abortus* in the United States [7]. Transmission of *Brucella* spp. to susceptible cattle normally occurs by ingestion or oral contact with infected fetuses that have been aborted, fetal fluids and membranes, or uterine discharges [8]. Elk that congregate on feeding grounds from November through April overlap with the peak time period when *Brucella* is transmitted to other animals (February through June) [9]. Maichak et al. reported that as many as 12% of the elk attending feeding grounds come into contact with non-infectious elk fetuses placed on these sites [10]. Bison normally congregate in large numbers, which increases the likelihood they will come into contact with *Brucella*-infected fetuses and discharges. Such congregation increases the possibility that infected bison could transmit *Brucella* to cattle herds in the area [7]. As a result, farmers may unnecessarily kill elk or bison that wander out of the park and onto private farmlands.

Development of reliable and cost-effective diagnostic tests for use in elk and other wildlife is a high research priority. Reliable and portable diagnostic assays that can be carried out in the field by non-specialist personnel are urgently needed to minimize the spread of the disease among wildlife and its transmission to domestic animals and humans.

Biosensors combine biological molecules with a physicochemical transducer. Biological components incorporated into biosensors may include nucleic acids, enzymes, antibodies, etc., and the transducer may be optical, electrochemical, thermometric, or piezoelectric. Regardless, the detection of the target biological material results in a measurable signal. The advantages of optical fibers (light, inexpensive, and low interference) have established them as essential instruments of sensor technology [11]. Biosensors that consist of optical fibers transmit light based on total internal reflection through their transduction elements. The sensor produces a signal that can be analyzed and is in proportion to the concentration of the molecule that binds to the biological element on the sensor. Grating devices in the optical fiber induce a periodic variation in the refractive index of the optical fiber's core. As a result, there is a significant drop in the amount of light transmitted through the fiber at a specific wavelength. The specific wavelength can be changed to account for temperature, pressure, or the type of binding event [12].

Layer-by-layer films, also known as ionic self-assembled multilayer (ISAM) films, are a novel type of materials that enable the user to modify the structure and thickness of the thin film at nanometer levels. The assembly of such films is simple and inexpensive [13,14]. As a result, optical fibers containing these nanoscale overlays substantially enhance, through direct light transmission, the detection of antigen binding to antibody or DNA hybridizing to complementary DNA. Furthermore, these sensors can be organized into a device that is rugged and portable [15,16]. For the detection of infectious agents, these fiber-optic biosensors can be used as rapid diagnostic or screening tests prior to culture, serology, or other means of diagnosis. A variety of fiber grating-based biosensor platforms have recently been developed [17–20]. For the work described here, a nanomaterial optical fiber biosensor (NOFS) assay was successfully used to detect *Brucella* DNA in culture lysates and in infected animal tissues.

2. Materials and Methods

2.1. Oligonucleotide Primers and Probes

The oligonucleotide probes and primers (Table 1) were designed manually based on the DNA sequences of the respective genes/regions in GenBank and were purchased from Integrated DNA Technologies, Collinsville, IL, USA. The IS*711* DNA region is present in all known nomenspecies of *Brucella*, but not in other bacteria. Therefore, the primers IS*711*-For and IS*711*-Rev and probes IS*711*-BIO and IS*711*-DIG from the IS*711* region (Table 1) were used for detection of all *Brucella* nomenspecies and to distinguish them from other bacterial species. In order to identify and differentiate the three major *Brucella* nomenspecies (*Brucella abortus*, *B. melitensis*, and *B. suis*), the oligonucleotide probes BruAb2_0168 (GenBank accession AE017224.1), Melitensis_0466 (GenBank accession AE008918.1), and Suis_TraJ (GenBank accession CP024421.1) (Table 1) were used. A search using the NCBI BLAST program confirmed the specificity of the DNA regions used for identifying and distinguishing the respective *Brucella* species.

Table 1. Oligonucleotide probes and primers used for detecting major *Brucella* nomenspecies.

Probe/Primer Name	Sequence (5′ to 3′)	Comments
Probe-IS711-BIO	AAGCCAACACCCGGCCATTATGGT	The probe was biotinylated at the 3′ end
Probe-IS711-DIG	GGCCTACCGCTGCGAATA	The probe was labeled with digoxigenin at the 5′ end
Primer-IS711-For	TTGGCCTTGATCTGAGCCGT	
Primer-IS711-Rev	ATCGAAAGTCCACGCAGATG	
Probe-BruAb2_0168	TGGAACGACCTTTGCAGGCGAGATC	Biotin label at the 3′ end; specific for *B. abortus*
Probe_Melitensis_0466	CCAGCTTTTGGCCTTTTCCAGATTG	Biotin label at the 3′ end; specific for *B. melitensis*
Probe_Suis-TraJ	CCATGAGCGCCCGCATGTCCTCTTG	Biotin label at the 3′ end; specific for *B. suis*

2.2. Bacterial Strains and Culture Conditions Used

The *Brucella* nomenspecies and other bacterial species used as controls in this study are listed in Table 2. The bacteria were grown to mid-log phase in brain heart infusion (BHI) broth. Bacteria were harvested by centrifugation, washed, resuspended in phosphate buffered saline, pH 7.2 (PBS), and killed by boiling for 20 min (confirmed by viable plate count). Serial dilutions of killed cell suspensions were made in PBS, and genomic DNA was harvested using the DNAeasy Blood and Tissue kit (Qiagen, Valencia, CA, USA).

Table 2. Bacterial species and strains used.

Bacterial Species	Strain Name	Source
Brucella abortus	2308	Virginia Tech Biosafety 3 laboratory
Brucella melitensis	16 M	Virginia Tech Biosafety 3 laboratory
Brucella suis	1330	Virginia Tech Biosafety 3 laboratory
Klebsiella pneumoniae	2237	Virginia Tech Veterinary Teaching Hospital
Klebsiella pneumoniae	2237-13	Virginia Tech Veterinary Teaching Hospital
Proteus mirabilis	2172B	Virginia Tech Veterinary Teaching Hospital
Proteus mirabilis	13-2319	Virginia Tech Veterinary Teaching Hospital
Proteus mirabilis	13-2401	Virginia Tech Veterinary Teaching Hospital
Staphylococcus aures	29213	Virginia Tech Veterinary Teaching Hospital
Enterococcus faecalis	2174	Virginia Tech Veterinary Teaching Hospital
Enterococcus faecalis	13-2321-2	Virginia Tech Veterinary Teaching Hospital
Escherichia coli	2174	Virginia Tech Veterinary Teaching Hospital
Escherichia coli	Top10	ThermoFisher Scientific
Escherichia coli	13-2438	Virginia Tech Veterinary Teaching Hospital
Salmonella arizonae	13-2453	Virginia Tech Veterinary Teaching Hospital
Enterobacter aerogenes	13-2329-2	Virginia Tech Veterinary Teaching Hospital
Enterococcus faecium	13-2174-C	Virginia Tech Veterinary Teaching Hospital
Enterococcus faecium	13-2248-2	Virginia Tech Veterinary Teaching Hospital

2.3. PCR

PCR, when used, was performed in 25 µL volumes and included 10 pmol each of the primers Primer-IS711-For and Primer-IS711-Rev (Table 1), 1 mM $MgCl_2$, 200 µM dNTPS, 1X concentration of One*Taq* Standard Reaction Buffer (New England Biolabs, Ipswich, MA, USA), 1.25 units of One*Taq* DNA Polymerase (New England Biolabs), and template DNA. Template DNA included either 26 ng of genomic DNA or 1 µL of heat-killed, cell lysate from 1×10^5 cells/mL to 3×10^{10} cells/mL. Reaction conditions were an initial denaturation temperature of 95 °C for 5 min, followed by 30 cycles of 95 °C/1 min, 57 °C/1 min, 72 °C/1 min, and a final extension at 72 °C/10 min.

2.4. Enzyme-Linked Immunosorbent Assay (ELISA)

An ELISA was designed using Magnalink Streptavidin Magnetic beads (Solulink Inc., San Diego, CA, USA). The protocol was as described by the manufacturer (Solulink, Inc.) with modification.

Briefly, 60 pmol of the biotinylated probe (Probe-IS711-BIO) (Table 1) in 250 µL of nucleic acid binding and wash buffer (50 mM Tris-HCl, 150 mM NaCl, 0.05% Tween 20, pH 8.0) was incubated for 30 min at room temperature with the beads in 1.5 mL microcentrifuge tubes. The heat-denatured PCR products or genomic DNA were incubated with the beads coupled to the biotinylated probe in hybridization buffer (3× SSC, 0.05% Tween 20) for 2 h at 45 °C. A digoxigenin-labeled probe (Probe-IS711-DIG) (Table 1) in hybridization buffer was then incubated with the bead/probe/DNA triplex for 2 h at 45 °C. The DIG Detection Starter Kit from Roche (Sigma-Aldrich, St. Louis, MO, USA) was used to determine binding of the probe to the triplex complex. The ELISA was designed solely to confirm that the designed probe would hybridize with the *Brucella* genomic DNA, and not as a diagnostic assay itself. Therefore, quantitative data were not obtained.

2.5. Fabrication of the ISAM Film

ISAM films were fabricated using procedures described by the authors [21]. The ISAM method simply involves the alternate dipping of a charged substrate (optical fiber) into an aqueous solution of a polycation and an aqueous solution of a polyanion at room temperature. The optical fiber was immersed in an aqueous 10 mM poly-allylamine hydrochloride (PAH) (pH 7.0) solution for 2 min then rinsed three times in distilled water. The fiber was then immersed in a similar aqueous solution of 10 mM poly-1-[p-(3′-carboxy-4′-hydroxyphenylazo) benzenesulfonamido]-1,2-ethanediyl (PCBS) (pH 7.0) for 2 min and rinsed again. The final layer was always the negatively-charged PCBS. These two steps were repeated until the optimal number of bilayers was obtained, which, for this assay, was four layers (Figure 1).

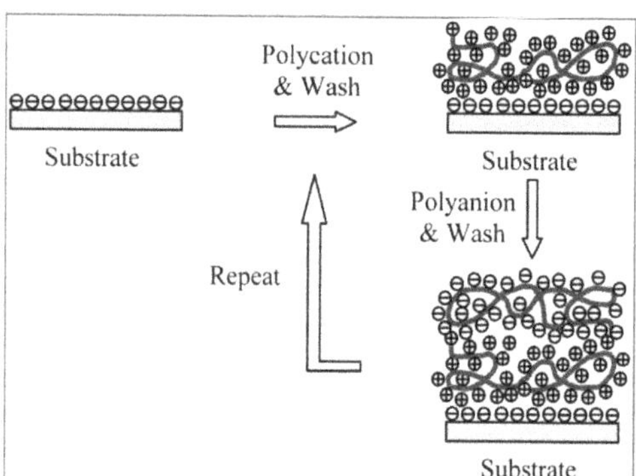

Figure 1. Assembly of the ionic self-assembled multilayer (ISAM) film. Polycationic and polyanionic solutions were alternately deposited on the optical fiber to form the ISAM film.

2.6. Coupling the Probe to the ISAM Film

The ISAM film was incubated with 0.6 mL of 0.17 M freshly prepared N-(3-dimethylaminodipropyl)-N′-ethylcarbodiimide (EDC), 0.17 M N-hydroxysulfosuccinimide (NHS), and 60 pmol of biotinylated oligonucleotide probe in PBS, pH 7.0, at room temperature for 30 min.

2.7. Conjugation of Streptavidin to the ISAM Film

An alternative method to couple the probe onto the ISAM film involved using a streptavidin intermediate. Four bilayers were deposited onto the optical fiber, leaving PCBS with negatively-charged carboxyl groups exposed. Then, 40 µL of streptavidin (1 mg/mL in PBS, pH 7.0) was mixed with 0.6 mL of cross-linker solution (0.17 M EDC and 0.17 M NHSS in PBS, pH 7.0). The mixture was added to the fiber and incubated for 8 h, with mixing every 15 min. The fiber was then rinsed and the biotinylated probe was added for spontaneous coupling to streptavidin.

2.8. NOFS Assay

The NOFS assay consists of turnaround point long-period gratings (TAP-LPGs) with ISAM films adsorbed on fiber cladding. The TAP-LPGs are TrueWave RS^{TM} (OFS) single-mode optical fibers with a grating period of 116 μm written by a 248 nm excimer laser through a chrome-plated mask. The grating couples to the $LP_{0,14}$ cladding mode of the fiber. White light model SLD-11OESL003 (FiberLabs, Inc. Fugimino-Shi, Saitama, Japan) was coupled to the optical fiber, and the spectra were measured by an optical spectrum analyzer (ANDO AQ6317) following the deposition of materials onto the TAP-LPG.

Bacteria grown in broth medium were harvested and washed in PBS. Serials dilutions of cultures were inoculated to agar medium to determine the colony forming units (CFU)/mL. *Brucella* cultures were lysed by boiling for 30 min. Loss of viability was confirmed by viable plate count before removing the bacteria from the biosafety level-3 laboratory. Prior to beginning the assay, preparations of genomic DNA, PCR products, and lysates of bacterial cells were boiled for 5 min. The film/probe duplex was incubated with the heat-denatured sample (genomic DNA, DNA regions amplified by PCR, amplified DNA from killed cells, dilutions of lysed cells, or dilutions of extracts of tissues from infected animals) for 50 min to allow hybridization between the probe and sample DNA to occur [21]. The TAP-LPG was tuned beyond the turnaround point such that the two narrow-band peaks merged into a single broadband peak that changed attenuation strength as the coupling between the core and cladding mode was modified by the addition of material to the cladding surface. As light in the range of 1400–1700 nm was transmitted through the ISAM fiber, an optical analyzer was used to record the attenuation in light transmission at 1550 nm. This attenuation in light transmission occurred due to the increase in coupling of light out of the core of the optical fiber due to sample DNA hybridizing to the DNA probe. An example of the series of spectra, as material binds to the cladding surface, is shown in Figure 2. The thickness of the ISAM films used is determined in order to set the attenuation at approximately half of the maximum attenuation that occurs before peak split into two separate narrowband peaks.

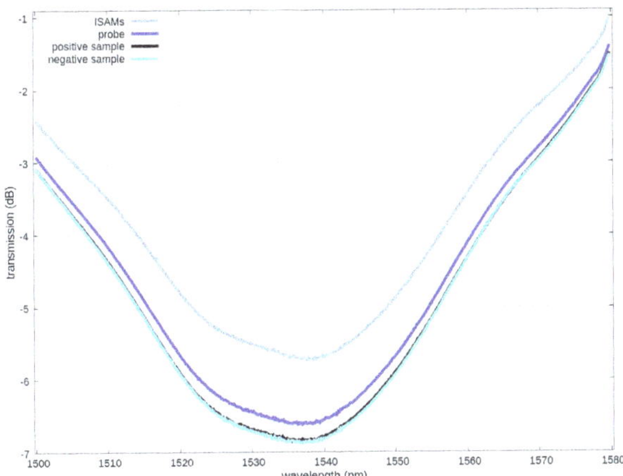

Figure 2. Transmission spectra after different steps of the assay. Adding the probe to the ISAM-coated fiber caused a large increase in attenuation. The attenuation was further increased after exposing the sensor to the positive control. However, further exposure of the fiber to the negative control did not result in any further change in attenuation, as was expected. As a result, the negative control spectrum overlaps and is indistinguishable from that of the positive control.

2.9. Detection of Brucella DNA in Tissues of Infected Mice

Groups of two female BALB/c mice each were inoculated intraperitoneally with 6×10^4 CFU/mouse of *B. abortus* strain 2308, *B. melitensis* strain 16 M, or *B. suis* strain 1330. Two mice were injected with PBS as controls. One week after inoculation the mice were euthanized, and 0.1 g of spleen and liver samples were collected. The tissues were ground with 1 mL of PBS. Half of the volume of the extracts of the ground tissues were used in viable plate count determination to determine the number of bacteria/g of tissue. The remaining half (500 µL of heat-denatured cell-free extract corresponding to 0.05 g of tissue) was used per each run of the NOFS assay, as previously described [21]. DNA in these samples were not amplified by PCR prior to NOFS testing. Serial dilutions of the extract were also cultured onto BHI agar, and bacterial colony counts were determined as CFU after 72 h of incubation at 37 °C with 5% CO_2.

2.10. Statistical Analyses

The standard deviations of the means were calculated from assays repeated at least three times. The online calculator (http://www.danielsoper.com/statcalc3/calc.aspx?id=43) was used to determine the analysis of variance, which was used to compare the transmission attenuation between different samples. The online calculator (http://www.socscistatistics.com/tests/studentttest/Default2.aspx) was used to calculate *p*-values from the Student *t*-test and to compare the attenuation of light transmission recorded for infected versus control tissue extracts. Student *t*-tests were also used for analysis of the attenuation of light transmission after exposure of the probe to two different PCR products. Results with calculated *p*-values of less than 0.05 was considered significant. The cutoff value in percent attenuation of light transmission that was used to differentiate negative from positive samples was calculated by multiplying the standard deviation of the true negative isolates tested by 3. This cutoff value could change depending on the optical fiber used and varied from 0.6% light attenuation to 3.2% light attenuation. Larger cutoff values were due to larger standard deviations of the negative controls.

3. Results

3.1. Specificity of the DNA Probes

DNA amplification of *Brucella* and heterologous species using oligonucleotide primers to the IS*711* region (Table 1) confirmed the specificity of the IS*711* region for *Brucella* nomenspecies. An approximately 300 bp-size amplicon was obtained when 50 ng of genomic DNA or lysates containing at least 5×10^3 killed cells of each *Brucella* nomenspecies was used in PCR reactions. However, visible amplicons were not seen in agarose gels when lysates representing 8×10^2 or fewer *Brucella* cells were used in PCR reactions. PCR amplicons were also not seen when lysates containing up to 3×10^7 killed cells of *Escherichia coli*, *Pseudomonas aeruginosa*, or *Salmonella* Typhimurium (negative controls) were used (Figure 3).

3.2. Validation of Target DNA for Hybridization to the DNA Probe

An ELISA was used to validate that target DNA hybridized to probes of the IS*711* region. After the DNA and initial bead-bound probe were allowed to hybridize, a second DIG-labeled oligonucleotide IS*711* probe to a different region of the DNA was added. Only if the sample DNA bound to the first probe would the second DIG-labelled probe bind and specifically detect *Brucella* DNA. The use of genomic DNA or lysate of killed cells in the absence of PCR did not produce a colorimetric change, indicating there was inadequate complementary DNA sequence from the genomic DNA to be detected in this assay. However, following DNA amplification of the test sample (genomic DNA or lysate containing 8×10^3 cells of *Brucella*), a positive reaction was obtained (Figure 4), but not if lysate representing 8×10^2 or fewer *Brucella* cells were tested (not shown). These results were consistent with results obtained by gel electrophoresis of PCR products and confirmed that the probe successfully bound to amplified DNA from the IS*711* region and was valid for use in the NOFS assay.

Figure 3. PCR amplicons from *Brucella* and control strains. Lanes and lysates representing the number of cells from nomenspecies used for PCR: 1 and 21, molecular size standards; 5 and 11, blank wells; 2–4, 3×10^7, 3×10^5, and 3×10^3 cells of *Pseudomonas aeruginosa* (the same negative results for the same number of cells are not shown for *Escherichia coli* and *Salmonella* Typhimurium); 6–10, 8×10^6, 8×10^5, 8×10^4, 8×10^3, and 8×10^2 cells of *B. abortus*; 12–16, 6×10^6, 6×10^5, 6×10^4, 6×10^3, and 6×10^2 cells of *B. suis*; 17–20, 5×10^6, 5×10^5, 5×10^4, and 5×10^3 cells of *B. melitensis*; 22, positive control (26 ng of *B. abortus* genomic DNA; 23, negative control.

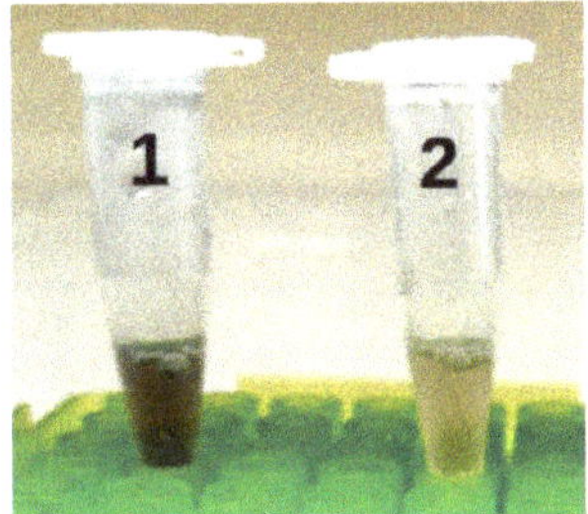

Figure 4. Magnetic bead ELISA. All tubes contained the beads, biotinylated probe of IS*711* gene, and digoxigenin-labelled probe to a distinct IS*711* DNA region. Tube 1 contained the PCR amplicon from a lysate of 7×10^6 cells of *Brucella abortus*. Tube 2 contained all the reaction components of tube 1, but the genomic DNA was not amplified by PCR.

3.3. Identification of Brucella Nomenspecies by NOFS Assay

Reaction of the ISAM/probe (IS*711*) duplex with the entire 25 µL of PCR amplicons from a lysate representing 10^4 cells of *B. abortus* strain 2308, *B. melitensis* strain 16 M, or *B. suis* strain 1330 in 500 µL of PBS resulted in 18.7%, 18.6% and 20.11% attenuation of light transmission, respectively, with positive results being above 0.6% light attenuation. When lysate from 10^2 cells of these same nomenspecies were tested, 8.8%, 14.2% and 13.6% attenuation of light transmission was obtained, respectively (Figure 5). These results indicated that the NOFS assay was capable of detecting PCR products with at least 10^2 cells of *Brucella*, which is much lower than the number of cells that could be detected by gel electrophoresis or ELISA. In contrast, when lysate from 10^4 cells of *P. aeruginosa*, *E. coli*, or *S.* Typhimurium were tested by the NOFS assay following PCR, less than 0.2% attenuation of transmission was obtained for any of the non-*Brucella* species tested (Figure 5).

Reaction of the ISAM/IS*711* probe duplex containing streptavidin with lysate representing 4×10^2 or 4×10^4 cells of heat-killed *B. abortus* without the use of PCR resulted in 4.3% and 14.5% transmission attenuation, respectively. Reaction of the same duplex lysate representing 5×10^4 cells of heat-killed *E. coli* failed to produce a positive transmission attenuation (Figure 6). These results confirmed that the assay could specifically detect low numbers of *Brucella* without the use of PCR.

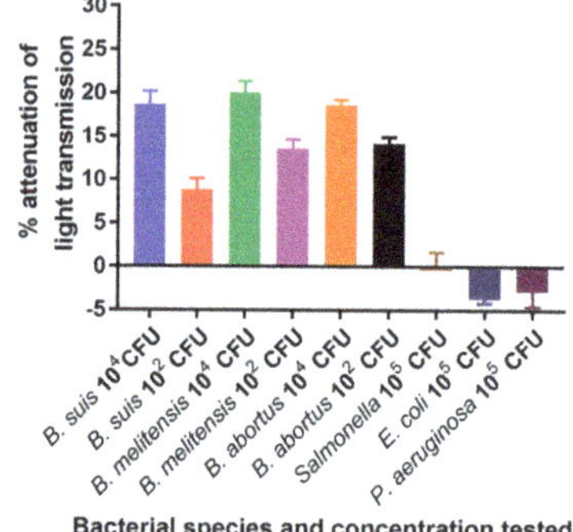

Figure 5. Detection of *Brucella* DNA amplified by PCR from *B. suis*, *B. abortus*, and *B. melitensis* by nanomaterial optical fiber biosensors (NOFS) assay. Each experiment consisted of 3 sequential steps: The biosensor was first tested with sample amplified by PCR from a lysate containing 10^4 cells of a negative control strain (*E. coli*, *Salmonella*, or *P. aeruginosa*), followed by an amplified sample of lysate from 10^4 *Brucella* cells, then lysate from an amplified *Brucella* culture containing 10^2 cells.

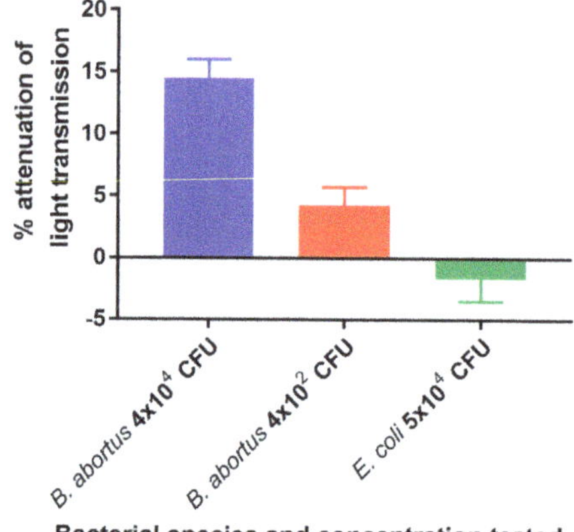

Figure 6. Detection of *B. abortus* DNA from lysates of killed cells by NOFS assay without PCR amplification. This experiment consisted of 3 sequential steps: The biosensor was first tested with lysate representing 5×10^4 cells of a negative control strain (*E. coli*), followed by lysate containing 4×10^4 cells of *B. abortus*, followed by lysate containing 4×10^2 cells of *B. abortus*.

When 10 replicates of each of the *Brucella* nomenspecies above were tested with lysates containing 10^2 cells/reaction by NOFS with streptavidin and without PCR, all were positive for attenuation of light transmission and significantly greater in comparison to three different non-*Brucella* species tested in duplicate as negative controls ($p \leq 0.0004$, pooled averages). The average light attenuation for *B. abortus* was 3.81% ± 0.92%, for *B. suis* was 3.50% ± 1.15%, and for *B. melitensis* was 5.15% ± 1.63% (all above the respective cutoff value for a positive result). Light attenuation results using lysates containing 10^4 cells/reaction of the negative control species *E. coli*, *P. aeruginosa*, and *Salmonella* were 0.41% ± 1.28%, 0.93% ± 1.68%, and 1.04% ± 0.89%. These results could be obtained in 30 min and confirmed that the NOFS assay was a highly sensitive, specific, and rapid assay for the detection of *Brucella* DNA.

3.4. NOFS Assay to Detect and Distinguish Different Brucella Nomenspecies and to Distingusih Brucella from Non-Brucella Bacterial Types

When the ISAM/probe BruAb2_0168 complex (specific for *B. abortus* but not for other *Brucella* nomenspecies) was reacted directly with lysate containing 10^5 heat-killed cells of *B. abortus* strain 2308 (without PCR), light transmission was attenuated by 5.4%. However, when the same ISAM/probe complex was reacted with a similar number of *B. melitensis* 16 M cells, transmission was attenuated only by 0.2%. In a separate assay, when the ISAM/probe Suis_*TraJ* complex (specific for *B. suis*) was reacted with lysate containing 10^5 cells of *B. suis* strain 1330, light transmission was attenuated by 3.8%. However, when the same ISAM/probe complex was reacted with lysate containing 10^5 cells of *B. abortus*, no positive attenuation of transmission was observed. When the ISAM/probe IS*711* complex was reacted with lysate representing 10^5 cells of 15 non-*Brucella* bacterial samples (Table 3), less than 2.2% light transmission attenuation was observed (all below the respective positive cutoff value). Thus, the NOFS assay was specific for *Brucella* and could detect and distinguish different *Brucella* nomenspecies.

Table 3. Percent NOFS light attenuation from non-*Brucella* bacterial types.

Bacterial Species	Strain Name	Transmission Attenuation (Mean ± Standard Deviation)
Klebsiella pneumoniae	2237	$-1.67\% \pm 0.15\%$
Klebsiella pneumoniae	2237-13	$2.09\% \pm 2.63\%$
Proteus mirabilis	2172B	$0.65\% \pm 0.31\%$
Proteus mirabilis	13-2319	$0.70\% \pm 1.64\%$
Proteus mirabilis	13-2401	$0.25\% \pm 0.95\%$
Staphylococcus aures	29213	$1.30\% \pm 0.70\%$
Enterococcus faecalis	2174	$1.57\% \pm 1.03\%$
Enterococcus faecalis	13-2321-2	$1.04\% \pm 0.85\%$
Escherichia coli	2174	$0.23\% \pm 0.06\%$
Escherichia coli	Top10	$-0.62\% \pm 0.94\%$
Escherichia coli	13-2438	$1.54\% \pm 0.12\%$
Salmonella arizonae	13-2453	$-0.59\% \pm 1.27\%$
Enterobacter aerogenes	13-2329-2	$1.64\% \pm 0.46\%$
Enterococcus faecium	13-2174-C	$2.14\% \pm 0.10\%$
Enterococcus faecium	13-2248-2	$1.69\% \pm 0.38\%$

3.5. Identification of Brucella in Tissues from Infected Mice by NOFS Assay

When the ISAM/probe BruAb2_0168 complex (specific for *B. abortus*) was reacted with 2 spleen or 2 liver extracts from *B. abortus*-infected mice, light transmission was attenuated by 6.79% ± 0.34% and 3.38% ± 0.78%, respectively (positive values were above 3.2% light attenuation for these assays). The average bacterial loads in the spleen and liver extracts used in the assays were 3.8×10^4 and 4×10^3 cells, respectively. However, when the same ISAM/probe complex was reacted with 2 spleen or 2 liver extracts from control mice inoculated with PBS, there was no positive attenuation of transmission for any of the samples (mean attenuation was −1.47% and −1.78%, respectively). Therefore, the NOFS assay could detect *B. abortus* in infected mouse spleen and liver. When the ISAM/probe Melitensis_0466 (specific for *B. melitensis*) was reacted with 2 spleen or 2 liver extracts from *B. melitensis*-infected mice, light transmission was attenuated by 7.6% and 6.1%, respectively. However, when the same ISAM/probe complex was reacted with 2 spleen or 2 liver extracts from PBS-injected mice, positive attenuation of transmission was not seen. Similarly, when the ISAM/probe Suis_TraJ complex (specific for *B. suis*) was reacted with 2 spleen or 2 liver extracts from *B. suis*-infected mice, light transmission was attenuated by 6.9% and 5.1%, respectively, but light transmission was less than 1% when the probe complex was reacted with 2 spleen or 2 liver extracts from PBS-injected mice (Table 4).

Table 4. Percent NOFS [a] light attenuation from spleen and liver extracts from mice infected with *Brucella* spp.

Challenge	Organ	Average % Attenuation ± Standard Deviation	Test Result [b]
With probe BruAb2_0168 (specific for *B. abortus*) *B. abortus*	Spleen (from mice 1 and 2)	6.80 ± 0.32	Positive
	Liver (from mice 1 and 2)	3.38 ± 0.78	Positive
PBS	Spleen (from mice 7 and 8)	−0.74 ± 1.12	Negative
	Liver (from mice 7 and 8)	−1.78 ± 0.29	Negative
With probe Melitensis_0466 (specific for *B. melitensis*) *B. melitensis*	Spleen (from mice 3 and 4)	7.60 ± 1.86	Positive
	Liver (from mice 3 and 4)	6.13 ± 2.81	Positive
PBS	Spleen (from mice 7 and 8)	−1.78 ± 2.60	Negative
	Liver (from mice 7 and 8)	−0.64 ± 0.41	Negative
With probe Suis_TraJ (specific for *B. suis*) *B. suis*	Spleen (from mice 5 and 6)	6.94 ± 0.59	Positive
	Liver (from mice 5 and 6)	5.10 ± 1.45	Positive
PBS	Spleen (from mice 7 and 8)	0.72 ± 2.93	Negative
	Liver (from mice 7 and 8)	0.34 ± 1.59	Negative

[a] All samples included streptavidin as a linker in the NOFS assay. [b] The positive cutoff value for this assay was 3.2%.

4. Discussion

Biosensors are becoming established diagnostic modalities and, when combined with PCR, have been used for detection of DNA using impedance spectroscopy [22,23] and piezoelectric gold electrode [24]. However, such biosensors are very expensive (i.e., may cost over $10,000 apiece), require high maintenance by experienced personnel, and have the additional PCR step. Therefore, these assays may also not be practical for many laboratories. Optical transduction methods such as surface plasmon resonance (SPR) are rapid and sensitive devices that have been developed for detection of bacterial agents [25]. However, these assays require the use of LED and spectroscopy to generate excited light and receive a signal. SPR sensors are also expensive and require highly trained personnel. Unlike other published biosensors, the NOFS assay described here utilizes nanometer-thick layers that can include a variety of materials, such as DNA, antibodies, and antigens. TAP-LPGs that are coupled to a DNA probe specific to the bacterium and supplemented with additional layers of biotin-streptavidin further enhance the limit of detection of the assay. As a result, PCR was not required for adequate detection of DNA with this NOFS assay. When the target DNA binds to the complementary oligonucleotide probe, thus altering the thickness of the film, the refractive index is also changed. As a result, the transmission characteristics of the fiber are modified, resulting in attenuation of the percent light transmitted. Due to the high specificity of the compatible DNA probe and target, specific DNA can be detected.

DNA probe assays have previously been used to identify the common nomenspecies of *Brucella* [26]. Specific primers have been used with DNA amplification to successfully differentiate *B. abortus* biovars 1, 2, and 4, *B. melitensis*, *B. suis* biovar 1, and *B. ovis*. [27,28]. Several investigators have shown that by targeting highly conserved genes (i.e., 16S rRNA [29], 16S-23S intergenic spacer regions [30], *bcsp*31 or IS*711* for all *Brucella* species [31,32], *alk*B for *B. abortus*, and BMEI1162 for *B. melitensis* [27,33]), probes and primers can be developed for direct detection of these agents. In this communication, DNA amplification was used to confirm that the IS*711* DNA region was specific for all three nomenspecies of *Brucella*, and an ELISA was used to demonstrate that the oligonucleotide probe specifically binds to the IS*711* region of *Brucella*. The NOFS assays, which included a biotin-streptavidin linker, were used to detect as few as 100 cells of *Brucella* with a high degree of sensitivity and specificity in the set of samples studied here, even without prior PCR amplification. The NOFS assay was also capable of detecting *Brucella* in the tissues of infected mice. The probes to BruAb2_0168, Melitensis_0466, and Suis_TraJ DNA regions were specific for *B. abortus*, *B. melitensis*, and *B. suis*, respectively, as determined by NOFS. Therefore, these designed oligonucleotide probes could be used to distinguish each of the respective *Brucella* nomenspecies from each other or heterologous bacteria. Major advantages of the NOFS assay were that it could be completed in less than 1 h, did not require particular expertise to perform, and did not require a large amount of bench space.

The detection of antibodies to the lipopolysaccharide (LPS) O-antigen by serological methods is the accepted diagnostic method for brucellosis in all hosts. However, specificity can be problematic due to the structural similarity of the O-antigen side chain of *Brucella* with that of other bacteria, particularly *Yersinia enterocolitica* O:9, *Vibrio cholerae*, and *E. coli* 0:157. At this time, no other antigens have been identified that can successfully replace the LPS O-antigen in diagnostic assays. Molecular diagnostic tests are now important methods in clinical microbiology, although they remain restricted to larger laboratories that have the funds, expertise, and equipment to utilize this technology. Real-time (q)PCR assays can detect the DNA of infectious disease agents with same day results [34–36]. However, qPCR technology is restrictive due to the large cost of equipment and expertise needed to carry out these assays. Therefore, qPCR is normally not available in medical settings that have or utilize small laboratories, particularly in rural communities where infections due to *Brucella* may be more prevalent. Furthermore, *Brucella* can be exceptionally difficult to detect in blood, and although isolation from animal tissues may be more productive, as we show here by detection of *Brucella* by NOFS from mouse tissues, such isolation is normally not practical with human tissues. Nonetheless, the NOFS assay can be modified to detect antibodies to *Brucella* by coupling the antigen to the fiber, rather than a DNA probe, or alternatively coupling antibodies to the fiber to detect a specific antigen. We have described such an assay using antibody-coupled sensors to detect methicillin-resistant *Staphylococcus aureus* [21] and *Francisella tularensis* [37].

The most prominent reservoir of *B. abortus* in the United States in in bison and elk in the Yellowstone National Park area [7]. Cattle farmers are particularly concerned that bison or elk that wander onto their farmlands may be infected with *Brucella* and transmit the agent to their cattle, resulting in their loss of *Brucella*-free status. The NOFS assay has the advantage that samples collected from anesthetized animals or aborted fetuses could be used in a small regional facility to rapidly detect the presence of *B. abortus*. In addition, *B. suis* is the most prevalent *Brucella* nomenspecies in the United States and is present in feral hogs in 14 U.S. states [38]. *B. suis* can be transmitted to humans through hunting (field dressing and butchering) or other close contact [39]. Although *Brucella* diagnosis in humans can be difficult due to non-specific flu-like symptoms, detection of the agent in the animal's tissues can strongly support the diagnosis. Therefore, with the correct primers and probes, this NOFS assay can be adapted to detect any of the *Brucella* nomenspecies.

The NOFS assay described here is at the proof-of-principle stage. Further work will be required to develop a diagnostic test ready for regulatory approval. Such work will require a large number of *Brucella* strains of the different nomenspecies, as well as additional negative control strains, to adequately determine the sensitivity and specificity of the assay. Although the limit of detection of the assay has been determined (about 100 cells/mL), due to the low number of strains of *Brucella* available to us, sensitivity (defined as the true positive rate divided by false positives) was not able to be accurately calculated. Although the specificity of the assay appeared to be 100%, additional control strains encompassing a wide variety of species would need to be tested to confirm this. In addition, the NOFS assay can easily be modified to detect antigens, antibodies, or DNA from a wide variety of infectious agents, including viruses, fungi, and parasites, as well as other bacteria. In addition, the assay could be used to detect DNA encoding for antibiotic resistance genes to aid in screening patients that may be colonized with bacteria carrying specific antibiotic resistance genes. Such an assay would be highly beneficial in hospital infection control situations, particularly with an assay that can be completed in a short period of time for reasonable cost, and without the need for highly trained personnel.

Author Contributions: Project administration: T.J.I. and J.R.H.; investigation, A.B.B., Z.Z., and K.M.; formal analysis: A.B.B., Z.Z., and K.M.; writing—original draft preparation, A.B.B. and K.M.; writing—review and editing, T.J.I., J.R.H., A.B.B., and K.M.

Funding: This study was funded, in part, by USDA-NIFA award 2010-37610-30877.

Acknowledgments: We thank Ben Fox for excellent technical assistance.

Conflicts of Interest: The authors declare no conflict of interest.

References

1. Corbel, M.J. International committee on systematic bacteriology subcommittee on the taxonomy of *Brucella*—Report of the Meeting, 5 September 1986, Manchester, England. *Int. J. Syst. Bacteriol.* **1988**, *38*, 450–452. [CrossRef]

2. Mercier, E.; JumasBilak, E.; AllardetServent, A.; Ocallahan, D.; Ramuz, M. Polymorphism in *Brucella* strains detected by studying distribution of two short repetitive DNA elements. *J. Clin. Microbiol.* **1996**, *34*, 1299–1302.

3. Jahans, K.L.; Foster, G.; Broughton, E.S. The characterisation of *Brucella* strains isolated from marine mammals. *Vet. Microbiol.* **1997**, *57*, 373–382. [CrossRef]

4. Verger, J.M.; Grayon, M.; Cloeckaert, A.; Lefevre, M.; Ageron, E.; Grimont, F. Classification of *Brucella* strains isolated from marine mammals using DNA-DNA hybridization and ribotyping. *Res. Microbiol.* **2000**, *151*, 797–799. [CrossRef]

5. Buchanan, T.M.; Hendrick, S.L.; Patton, C.M.; Feldman, R.A. Brucellosis in United States, 1960–1972—Abattoir associated disease. 3. Epidemiology and evidence for acquired immunity. *Medicine* **1974**, *53*, 427–439. [CrossRef] [PubMed]

6. Young, E.J. Human Brucellosis. *Rev. Infect. Dis.* **1983**, *5*, 821–842. [CrossRef]

7. Cheville, N.F.; McCullough, D.R.; Paulson, L.R. Brucellosis in the Greater Yellowstone area. National Research Council. In *Brucellosis in the Greater Yellowstone Area*; The National Acadimies Press: Washington, DC, USA, 1998.

8. Thorne, E.T. Brucellosis. In *Infectious Diseases of Wild Mammals*, 3rd ed.; Williams, E.S., Barker, I.K., Eds.; Wiley Online Library: Hoboken, NJ, USA, 2001; pp. 72–395.

9. Cross, P.C.; Edwards, W.H.; Scurlock, B.M.; Maichak, E.J.; Rogerson, J.D. Effects of management and climate on elk brucellosis in the Greater Yellowstone Ecosystem. *Ecol. Appl.* **2007**, *17*, 957–964. [CrossRef] [PubMed]

10. Maichak, E.J.; Scurlock, B.M.; Rogerson, J.D.; Meadows, L.L.; Barbknecht, A.E.; Edwards, W.H.; Cross, P.C. Effects of management, behavior, and scavenging on risk of brucellosis transmission ifn Elk of Western Wyoming. *J. Wildl. Dis.* **2009**, *45*, 398–410. [CrossRef]

11. Ramsden, J.J. Optical biosensors. *J. Mol. Recognit.* **1997**, *10*, 109–120. [CrossRef]

12. Kersey, A.D.; Davis, M.A.; Patrick, H.J.; LeBlanc, M.; Koo, K.P.; Askins, C.G.; Putnam, M.A.; Friebele, E.J. Fiber Grating Sensors. *J. Lightwave Technol.* **1997**, *15*, 1442–1463. [CrossRef]

13. Decher, G. Fuzzy nanoassemblies. *Science* **1997**, *277*, 1232–1237. [CrossRef]

14. Decher, G.H.J.D.; Hong, J.D.; Schmitt, J. Buildup of ultrathin multilayer films by a self-assembly process: Consecutively alternating adsorption of anionic and cationic polyelectrolytes on charged surfaces. *Thin Solid Film.* **1992**, *210/211*, 831–835. [CrossRef]

15. Wang, Z.; Heflin, J.R.; Van Cott, K.; Stolen, R.H.; Ramachandran, S.; Ghalmi, S. Biosensors employing ionic self-assembled multilayers adsorbed on long-period fiber gratings. *Sens. Actuators B Chem* **2009**, *139*, 618–623. [CrossRef]

16. Wang, Z.; Heflin, J.; Stolen, R.; Ramachandran, S. Analysis of optical response of long period fiber gratings to nm-thick thin-film coating. *Opt. Express* **2005**, *13*, 2808–2813. [CrossRef]

17. Bandara, A.B.; Zuo, Z.; Ramachandran, S.; Ritter, A.; Heflin, J.R.; Inzana, T.J. Detection of methicillin-resistant staphylococci by biosensor assay consisting of nanoscale films on optical fiber long-period gratings. *Biosens. Bioelectron.* **2015**, *70*, 433–440. [CrossRef] [PubMed]

18. Chiavaioli, F.; Baldini, F.; Tombelli, S.; Trono, C.; Giannetti, A. Biosensing with optical fiber gratings. *Nanophotonics* **2017**, *6*, 663–679. [CrossRef]

19. Esposito, F.; Sansone, L.; Taddei, C.; Campopiano, S.; Giordano, M.; Iadicicco, A. Ultrasensitive biosensor based on long period grating coated with polycarbonate-graphene oxide multilayer. *Sens. Actuators B Chem.* **2018**, *274*, 517–526. [CrossRef]

20. Piestrzyńska, M.; Dominik, M.; Kosiel, K.; Janczuk-Richter, M.; Szot-Karpińska, K.; Brzozowska, E.; Shao, L.; Niedziółka-Jonsson, J.; Bock, W.J.; Śmietana, M. Ultrasensitive tantalum oxide nano-coated long-period gratings for detection of various biological targets. *Biosens. Bioelectron.* **2019**, *133*, 8–15. [CrossRef]

21. Bandara, A.B.; Zuo, Z.; McCutcheon, K.; Ramachandran, S.; Heflin, J.R.; Inzana, T.J. Identification of *Histophilus somni* by a nanomaterial optical fiber biosensor assay. *J. Vet. Diagn. Investig.* **2018**, *30*, 821–829. [CrossRef] [PubMed]

22. Corrigan, D.K.; Schulze, H.; Henihan, G.; Ciani, I.; Giraud, G.; Terry, J.G.; Walton, A.J.; Pethig, R.; Ghazal, P.; Crain, J.; et al. Impedimetric detection of single-stranded PCR products derived from methicillin resistant *Staphylococcus aureus* (MRSA) isolates. *Biosens. Bioelectron.* **2012**, *34*, 178–184. [CrossRef]

23. Wang, Z.; Zhang, J.; Chen, P.; Zhou, X.; Yang, Y.; Wu, S.; Niu, L.; Han, Y.; Wang, L.; Chen, P.; et al. Label-free, electrochemical detection of methicillin-resistant *Staphylococcus aureus* DNA with reduced graphene oxide-modified electrodes. *Biosens. Bioelectron.* **2011**, *26*, 3883–3886. [CrossRef]

24. Tombelli, S.; Minunni, M.; Santucci, A.; Spiriti, M.M.; Mascini, M. A DNA-based piezoelectric biosensor: Strategies for coupling nucleic acids to piezoelectric devices. *Talanta* **2006**, *68*, 806–812. [CrossRef]

25. Yang, K.L.A.; Wu, S.Y.; Kwok, H.C.; Ho, H.P.; Kong, S.K. Using loop-mediated isothermal DNA amplification (LAMP) and spectral surface plasmon resonance (SPR) to detect methicillin-resistance *S. aureus* (MRSA). In Proceedings of the 2012 International Conference on Biomedical Engineering and Biotechnology, Macao, China, 28–30 May 2012. [CrossRef]

26. Gotuzzo, E.; Carrillo, C.; Guerra, J.; Llosa, L. An evaluation of diagnostic methods for brucellosis—The value of bone marrow culture. *J. Infect. Dis.* **1986**, *153*, 122–125. [CrossRef] [PubMed]

27. Bricker, B.J.; Halling, S.M. Differentiation of Brucella abortus BV-1, Brucella abortus BV-2, and Brucella abortus BV-4, Brucella mellitensis, Brucella ovis, and Brucella suis BV-1 by PCR. *J. Clin. Microbiol.* **1994**, *32*, 2660–2666.

28. Matar, G.M.; Khneisser, I.A.; Abdelnoor, A.M. Rapid laboratory confirmation of human brucellosis by PCR analysis of a target sequence on the 31-kilodalton *Brucella* antigen DNA. *J. Clin. Microbiol.* **1996**, *34*, 477–478. [PubMed]

29. Wellinghausen, N.; Nockler, K.; Sigge, A.; Bartel, M.; Essig, A.; Poppert, S. Rapid detection of *Brucella* spp. in blood cultures by fluorescence in situ hybridization. *J. Clin. Microbiol.* **2006**, *44*, 1828–1830. [CrossRef]

30. Rijpens, N.P.; Jannes, G.; VanAsbroeck, M.; Rossau, R.; Herman, L.M.F. Direct detection of *Brucella* spp. in raw milk by PCR and reverse hybridization with 16S-23S rRNA spacer probes. *Appl. Environ. Microbiol.* **1996**, *62*, 1683–1688.

31. Bricker, B.J.; Ewalt, D.R.; Halling, S.M. *Brucella* 'HOOF-Prints': Strain typing by multi-locus analysis of variable number tandem repeats (VNTRs). *BMC Microbiol.* **2003**, *3*, 15. [CrossRef]

32. Ewalt, D.R.; Bricker, B.J. Validation of the abbreviated *Brucella* AMOS PCR as a rapid screening method for differentiation of *Brucella abortus* field strain isolates and the vaccine strains, 19 and RB51. *J. Clin. Microbiol.* **2000**, *38*, 3085–3086.

33. Probert, W.S.; Schrader, M.N.; Khuong, N.Y.; Bystrom, S.L.; Graves, M.H. Real-time multiplex PCR assay for detection of *Brucella* spp., *B. abortus*, and *B. melitensis*. *J. Clin. Microbiol.* **2004**, *42*, 1290–1293. [CrossRef] [PubMed]

34. Schmalz, G.; Tsigaras, S.; Rinke, S.; Kottmann, T.; Haak, R.; Ziebolz, D. Detection of five potentially periodontal pathogenic bacteria in peri-implant disease: A comparison of PCR and real-time PCR. *Diagn. Microbiol. Infect. Dis.* **2016**, *85*, 289–294. [CrossRef] [PubMed]

35. Thorn, M.; Rorsman, F.; Ronnblom, A.; Sangfelt, P.; Wanders, A.; Eriksson, B.M.; Bondeson, K. Active cytomegalovirus infection diagnosed by real-time PCR in patients with inflammatory bowel disease: A prospective, controlled observational study. *Scand. J. Gastroenterol.* **2016**, *51*, 1075–1080. [CrossRef] [PubMed]

36. Fellahi, S.; Harrak, M.E.; Kuhn, J.H.; Sebbar, G.; Bouaiti, E.A.; Khataby, K.; Fihri, O.F.; Houadfi, M.E.; Ennaji, M.M. Comparison of SYBR green I real-time RT-PCR with conventional agarose gel-based RT-PCR for the diagnosis of infectious bronchitis virus infection in chickens in Morocco. *BMC Res. Notes* **2016**, *9*, 231. [CrossRef]

37. Cooper, K.L.; Bandara, A.B.; Wang, Y.; Wang, A.; Inzana, T.J. Photonic biosensor assays to detect and distinguish subspecies of *Francisella tularensis*. *Sensors* **2011**, *11*, 3004–3019. [CrossRef]
38. eXtension. Brucellosis in Feral Hogs. Cooperative Extension Service. 2012. Available online: https://articles.extension.org/pages/65499/brucellosis-in-feral-hogs (accessed on 18 March 2019).
39. Giurgiutiu, D.; Banis, C.; Hunt, E.; Mincer, J.; Nicolardi, C.; Weltman, A.; Stanek, D.; Matthews, S.; Siegenthaler, C.; Blackmore, C.; Tiller, R.; et al. Centers for Disease Control. *Brucella suis* infection associated with feral swine hunting—Three States, 2007–2008. *Morb. Mortal. Wkly. Rep.* **2009**, *58*, 618–621.

MDPI

St. Alban-Anlage 66

4052 Basel

Switzerland

Tel. +41 61 683 77 34

Fax +41 61 302 89 18

www.mdpi.com

Biosensors Editorial Office

E-mail: biosensors@mdpi.com

www.mdpi.com/journal/biosensors